Leonardo Badia Thomas Marchioro

GAME THEORY

A handbook of problems and exercises

Società Editrice ESCULAPIO

ISBN 978-88-9385-286-9

Società Editrice Esculapio s.r.l.
Via Terracini, 30 - 40131 Bologna
www.editrice-esculapio.com - info@editrice-esculapio.it

Reprint 2022

A special thanks to Emilija Đokanović for her useful feedbacks.

Editorial staff: Carlotta Lenzi, Laura Tondelli, Laura Brugnoli

Società Editrice
ESCULAPIO

Contents

1 Static games of complete information 1

1.1 Characteristics 1
1.2 Best responses 2
1.3 Nash equilibria in pure strategies 3
1.4 Elimination of strictly dominated strategies 4
1.5 Nash equilibria in mixed strategies 4
Exercise 1.1 8
Exercise 1.2 10
Exercise 1.3 12
Exercise 1.4 14
Exercise 1.5 16
Exercise 1.6 18
Exercise 1.7 20
Exercise 1.8 22
Exercise 1.9 24
Exercise 1.10 25
Exercise 1.11 27

2 Zero-sum games and minimax 29

2.1 Characteristics 29
2.2 Maximin and minimax 30
Exercise 2.1 32
Exercise 2.2 34
Exercise 2.3 36
Exercise 2.4 38
Exercise 2.5 40

3 Sequential games 43

3.1 Characteristics 43
3.2 Backward induction 44
3.3 Subgame perfect equilibrium 46
Exercise 3.1 47
Exercise 3.2 49
Exercise 3.3 51
Exercise 3.4 53
Exercise 3.5 55
Exercise 3.6 57
Exercise 3.7 59
Exercise 3.8 61
Exercise 3.9 62

Exercise 3.10 64
Exercise 3.11 65

4 Lotteries and Other Variations **67**

4.1 Lotteries 67
4.2 Tampering with elections 68
4.3 Stackelberg games 69
Exercise 4.1 71
Exercise 4.2 73
Exercise 4.3 75
Exercise 4.4 77
Exercise 4.5 79
Exercise 4.6 80
Exercise 4.7 82
Exercise 4.8 84
Exercise 4.9 85
Exercise 4.10 86

5 Multistage Games **89**

5.1 Characteristics 89
5.2 Finite horizon 90
5.3 Infinitely repeated games 91
Exercise 5.1 93
Exercise 5.2 94
Exercise 5.3 96
Exercise 5.4 98
Exercise 5.5 100
Exercise 5.6 102
Exercise 5.7 104
Exercise 5.8 106
Exercise 5.9 107
Exercise 5.10 108

6 Bayesian games **109**

6.1 Characteristics 109
6.2 Relevant cases 111
6.3 Perfect Bayesian equilibrium 115
Exercise 6.1 118
Exercise 6.2 121
Exercise 6.3 123
Exercise 6.4 125
Exercise 6.5 128
Exercise 6.6 130
Exercise 6.7 132
Exercise 6.8 134
Exercise 6.9 136

Exercise 6.10 138
Exercise 6.11 140
Exercise 6.12 142
Exercise 6.13 144
Exercise 6.14 146
Exercise 6.15 148
Exercise 6.16 150
Exercise 6.17 153

1

Static games of complete information

Static games with complete information are the simplest form of games, in which players act simultaneously and without coordination. For example, rock-paper-scissors is one such game, since the players make their decisions simultaneously. Tic-tac-toe instead is not static, because moves unfold over time, and each player gets to see the opponent's previous move and can choose their next move accordingly. Complete information means that all players are aware of each other's payoff. Usually, these games give the first taste of concepts such as the Nash equilibrium, and are also important to introduce mixed strategies.

1.1 Characteristics

In a static game of complete information, the strategic choices of players are very simple. Each player must only select one of its available actions, and does so unbeknownst to the other players, so that there is no extra information available to exploit. This means that the "pure strategies" of the players are simply defined as their available actions. Generally, the scenario is considered with a finite set of actions, and formally speaking can be represented in normal form by defining

- a set of players, e.g., $\{\mathtt{P1}, \mathtt{P2}\}$;
- a set of strategies for each player, e.g., $\mathcal{S}_{\mathtt{P1}} = \{A, B, C\}$ and $\mathcal{S}_{\mathtt{P2}} = \{D, E\}$;
- payoffs for all players, e.g. $u_{\mathtt{P1}}(A, D) = 1$, $u_{\mathtt{P2}}(A, D) = 2$, $u_{\mathtt{P1}}(B, D) = 5$, $u_{\mathtt{P2}}(B, D) = 6$, etc.

However, writing the value of $u_{\mathtt{P1}}(s_1, s_2)$ for each pair of strategies is a bit tedious. It is more convenient and visually intuitive to adopt a matricial representation to enumerate the payoffs. Very often, 2-person games are considered, which makes this representation a bidimensional matrix (i.e., a real matrix instead of a hypercubic mess). Note, however, that each cell must contain 2 values, i.e., the payoffs of both players. An exception to this is represented by special cases such as zero-sum games, discussed later. For this reason, it would be more appropriate to call this a *bi-matrix*.

Notably, if your exercise involves more than two players (they do it sometimes) you should figure out a way to represent it properly. For example, a 3-person game where all players have 2 actions to choose from can still be represented with little effort, even on a bidimensional piece of paper, by writing one "floor" of the $2 \times 2 \times 2$ structure to the left and the other to the right. Keep in mind that each cell must contain 3 values, though.

For the most common case of games involving two players with a finite number of strategies, payoffs are written in a bi-matrix as follows:

P2

P1		D	E
	A	1, 2	3, 4
	B	5, 6	7, 8
	C	9, 10	11, 12

Each cell contains payoff pairs, the left one represents P1's utility, while the right one is P2's. In fact, the usual convention is that the first player chooses the row, and the second player chooses the column.

Example: A classic problem in Game Theory is known as the Prisoner's Dilemma. The original story of the problem relates to thieves being caught and interrogated by the police separately, and thus facing the choice on whether confess to the crime (C) or keep the mouth shut and wave Miranda rights (M). This game can be written in normal form with the following bi-matrix, where the payoff values represent the years spent in jail under that outcome (hence the negative sign: the more years, the worse).

P2

P1		M	C
	M	-1, -1	-9, 0
	C	0, -9	-6, -6

The Nash equilibrium of this game is that they both play C, despite this not being a particularly "good" outcome (especially, it is dominated by (M,M) in Pareto sense), hence the dilemma.

1.2 Best responses

Best responses are a fundamental concept to find Nash equilibria. A strategy s_1 of P1 is a best response to another strategy s_2 of P2 if s_1 is the one among P1's strategies giving highest payoff to P1 when P2 plays s_2. In other words, if there is no better strategy that P1 can play against s_2. A "trick" to see best responses is to look only at the payoffs of one player, covering those of the other players.

Example: Say we want to find best strategies for P1 in the Prisoner's dilemma. We can look only at his payoffs and check the best response for both strategies M and C of P2.

P2

P1		M	C
	M	-1, ·	-9, ·
	C	0, ·	-6, ·

The best response of P1 to M is to play C, since $0 > -1$. Similarly, C is also the best response of P1 to C, since $-6 > -9$. A common way of highlighting pure strategies that are best responses is to overline the best responses of P1 and to underline the best responses of P2, as done in the figure below.

P2

P1		M	C
	M	-1, -1	-9, 0
	C	0, -9	-6, -6

1.3 Nash equilibria in pure strategies

The Nash equilibrium is the key concept of game theory. A Nash equilibrium is a joint strategy (i.e., a tuple of strategy, one by each player) where all players choose a best response to the others' move. Intuitively, this means that each player will not regret the choice made, given the outcome of the game. Since Nash equilibria imply that everyone is playing a best response, those in pure strategies are found in a bi-matrix simply by inspecting which cells are both overlined and underlined. Nash equilibria in mixed strategies are explained in the next section.

Example: In the Prisoner's dilemma, (C, C) is the only joint pure strategy where both prisoners play a best response, which makes it also the only Nash equilibrium of the game. Notice that all the other joint strategies contain irrational behaviors. For instance, (M, M) is irrational because both players get in jail for 1 year. Both regret this choice, since by playing C they could have been set free immediately.

There can be games with multiple Nash equilibria in pure strategies, and games with none. Another classic problem, containing two Nash equilibria in pure strategies, is known as the Battle of Sexes. The colorful and originally not very politically correct explanation relates to two individuals who independently decide a spectacle to watch between options A and B. They have different valuations on what show is better, but their first and foremost goal is to watch it together, so the resulting payoff bi-matrix is written as

P2

P1		A	B
	A	2, 1	0, 0
	B	0, 0	1, 2

The two Nash equilibria are (A, A) and (B, B), meaning that the two players are happy if they end up together. An outcome like (A, B) is not rational as both would regret their choices. Scenarios like this are also known as *coordination games*.

1.4 Elimination of strictly dominated strategies

When solving exercises, it may also be useful to remember the concept of eliminating strictly dominated strategies. Formally, strategy S of player i is strictly dominated by T if $u_i(T, s_{-i}) > u_i(S, s_{-i})$, for all strategic choices s_{-i} of the other players (notation $-i$ is commonly used in game theory to denote the other players than player i). Also, note that the inequality must be strict and hold true for all possible s_{-i}'s.

Strictly dominated strategies are never chosen by rational players, thus if one such strategy S is spotted when solving an exercise, the corresponding row or column can be safely removed from the payoff matrix. Sometimes, this may lead to other strategies becoming strictly dominated too, and they can be iteratively eliminated, a process known as iterated elimination of strictly dominated strategies (or IESDS, for short).

While removing dominated strategies is not mandatory, it often simplifies the solution of an exercise to a great extent. Indeed, a strictly dominated strategy is never part of an NE and since the usual question is to find the NE, removing it from the game just leaves the solution with a smaller matrix that is easier to handle. For example, in the prisoner's dilemma, one can see that not only (C, C) is the Nash equilibrium (which can be proven even by just overlining-underlining best responses), but also M is a strictly dominated strategy (clearly, by C) for both players. Thus, (C, C) is the only available choice to rational players and is therefore the only NE.

Unfortunately, there is no standard procedure to spot strictly dominated strategies other than carefully check the payoff matrix and see if there is a row (or column) that always assigns to the related user an array of payoffs that is element-wise lower than another row (or column).

1.5 Nash equilibria in mixed strategies

So far, only pure strategies have been discussed. However, there are cases where it makes sense to consider mixed strategies, which involve randomness. A mixed strategy combining pure strategies $\{A, B, C\}$ (one may also say that the *support* of this mixed strategy consists of these three pure strategies) can be represented as $pA+qB+rC$, which means "play A with probability p, play B with probability q, play C with probability r". Since weights (p, q, r) form a probability distribution they must satisfy $p + q + r = 1$, i.e., $r = 1 - p - q$. Pure strategies can be reinterpreted as mixed strategies with a degenerate probability distribution in which only one value has probability 1.

Utilities in mixed strategies are just computed by taking expectations over the aforementioned probabilities. For example, in the Prisoner's dilemma, joint strategy $(0.9M + 0.1C, M)$ gives P2 a payoff equal to

$$u_{\mathrm{P2}}(0.9M + 0.1C, M) = 0.9u_{\mathrm{P2}}(M, M) + 0.1u_{\mathrm{P2}}(C, M) = 0.9 \cdot (-1) + 0.1 \cdot (-9) = -1.8.$$

The concept of strictly dominated strategies applies to mixed strategies too, both in the sense that mixed strategies can be strictly dominated and also they can strictly dominate another strategy, even a pure one. The latter case might be useful for solving exercises, but unfortunately spotting that a pure strategy is strictly dominated by a mixed strategy is generally not easy to do, since mixed strategies are not explicitly reported on the matrix, so it requires a bit of effort.

Nash equilibria in mixed strategies are found by applying the *indifference principle*. This states that a mixed strategy at the Nash equilibrium should give the same payoff to the other player for all strategies. That is, if P1 plays $pA + (1-p)B$ and P2 can choose between moves $\{D, E\}$, the condition to satisfy is

$$u_{\mathrm{P2}}(p, D) = u_{\mathrm{P2}}(p, E),$$

where p is short notation for $pA + (1-p)B$.

Example: In the game of "Odds and Evens" both players choose a number, usually between 0 and 5, and do so simultaneously (as in rock-paper-scissors). The number can indeed be odd (O) or even E. One player, say P1, wins and gets payoff $+1$ if the sum of the outcomes is even; then the other, P2, wins instead if the sum is odd. The loser gets payoff -1. Conveniently, it turns out that the sum relationships on the $\mathbb{Z}_2$ field, i.e., between odd and even numbers satisfy certain mathematical properties that allow to represent just the choice of O or E as the pure strategies. The normal form of the game is shown through the following bi-matrix

		P2	
		O	E
P1	O	1, -1	-1, 1
	E	-1, 1	1, -1

In general, games with a structure like that, i.e., one player likes when the selections are the same, but the other prefers that they are different, are called *discoordination games*. Compare this with a coordination game, where both players want instead them to be the same. A discoordination game does not have Nash equilibria in pure strategies, since all the joint strategies imply that one of the player has a regret and wants to change the outcome. However, an equilibrium in mixed strategies can be found by applying the indifference principle. P1 should play a mixed strategy $pO + (1-p)E$ such that P2's payoff is the same playing either O or E. This concept leads to

$$u_{\mathrm{P2}}(p, O) = u_{\mathrm{P2}}(p, E)$$

that is

$$pu_{P2}(O,O) + (1-p)u_{P2}(E,O) = pu_{P2}(O,E) + (1-p)u_{P2}(E,E)$$

Plugging in the numbers,

$$p(-1) + (1-p)(+1) = p(+1) + (1-p)(-1)$$

which is solved as $p = 1/2$

Likewise, P2 should play a mixed strategy $qO + (1-q)E$ such that P1's payoff is the same playing either strategies. By symmetry of the game, $q = p = 1/2$. This means that the best thing a player can do is to play even or odd with equal probability, which explains why the "Odds and Evens" game is used as a poor man's replacement for a fair coin flip (indeed, a very poor man if it is penniless).

Notice that for all $p < 1/2$, for P2 is more convenient to play E, while for all $p > 1/2$ O is P2's best response. P2's best response can thus be represented as a step function of p. Since the opposite applies to P1 (for $p < 1/2$ he should always play O, for $p > 1/2$ always E), a plot of both best responses looks as follows.

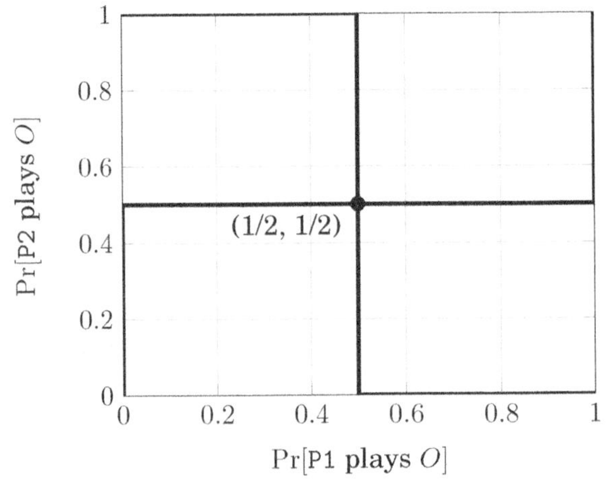

Figure 1.1: Mixed Nash equilibrium of the game.

Now, the most famous game theoretic result, Nash theorem, states the existence of at least one NE, possibly in mixed strategies, for each static game of complete information where the players have a finite set of available actions. Unfortunately, this is just an existence theorem, and does not say whether this NE is unique, or there are many. This implies that in certain situations the theorem can be promptly used to justify the existence of an NE in mixed strategy, for example, in discoordination games that are without NE in pure strategies, hence, they must have one in mixed strategies, but not to limit their number.

Also, for 2-person games where the sets of available actions is binary for both players, some standard structure can be recognized, such as games similar to the prisoner's dilemma, where one of the strategies is strictly dominated and therefore there is only one NE (in pure strategies), the result of IESDS. Or, coordination games, that can be shown to always have a mixed strategy NE in addition to the two on the diagonal of the payoff matrix. Moreover, when there are only 2 pure strategies to consider, say A and B, the indifference theorem is also simple to apply, since all non-degenerate mixed strategies must have a support consisting of A and B, indeed.

When the game has more than 2 players, or more than 2 pure strategies per player, it is not always easy to recognize a known structure instead, and very often it is required to do some tedious exploration of the strategy space. This is especially true for applying the indifference theorem, that can consider extra requirements. First of all, there are different combinations possible; such as out of three pure strategies A, B, and C , there are mixed strategies whose support is of just two strategies, such as A and B, or A and C, and also there are mixture where all strategies A, B, and C belong to the support.

Also remember that if, at an NE, a player chooses a mixed strategy that combines only A and B, it must not only hold that the player is indifferent between A and B, but also the resulting payoff must be not lower than what achieved when playing C, otherwise we would have a deviation toward another better strategy, i.e., C, which negates that this is an NE.

Exercises

Exercise 1.1. Jorah (J) and Khaleesi (K) are conquering cities in the continent of Essos. They are leading their armies to the cities on the coast, and they can choose between the enemy cities of Meeren (M) or Yunkai (Y); each of them has the opportunity to attack only one city. The inherent values of the two cities is $v_M = 5$ and $v_Y = 3$, respectively. If a city is attacked, even by just one army, victory is certain and the city is conquered. They separately decide what city to attack, and it is impossible for them to communicate in time what they decide. Also, while they both want to conquer the cities, their motives are different: K wants to conquer as many cities as possible, so her utility u_K is the sum of the values of the captured cities. While J has this in mind, he also wants to be closer to K, so his utility is the sum of the values of the captured city, plus an additional value of 10 if they choose the same city. All of this information (the values of the cities, and the crush that K has on J) is known to both J and K.

1. What kind of game is this? Write down its normal-form representation.
2. Find out all the Nash equilibria in pure strategies of this resulting game.
3. Find out all the Nash equilibria of this resulting game.

Solution

1. This is a static game of complete information. The normal-form representation can be shown as the following bi-matrix, where the set of players is $\{\mathtt{J}, \mathtt{K}\}$ and their strategy set is $\mathcal{S}_\mathtt{J} = \mathcal{S}_\mathtt{K} = \{M, Y\}$.

K

J		M	Y
	M	15, 5	8, 8
	Y	8, 8	13, 3

2. Nash equilibria are found when both J and K play a best response to the other's move. J's best responses to K's move in pure strategies are underlined, K's best responses are overlined.

K

J		M	Y
	M	$\underline{15}, 5$	$8, \overline{8}$
	Y	$8, \overline{8}$	$\underline{13}, 3$

The game has no Nash equilibria in pure strategies, since there is no joint pure strategy in which both J and K play a best response to the other's move.

3. Thus, there must exist a Nash equilibrium in mixed strategies. This can be found by setting a mixed strategy (p, q) where p and q the respective probabilities that J and K play strategy M. The values of p and q at the Nash equilibrium can be found through the indifference principle, since a mixed strategy contains both strategies M and Y in its support. Denoting simply with p the mixed strategy "play M with probability p and Y with probability $1-p$" (and same for q):

   - for player K, the condition to be satisfied is $u_J(M, q) = u_J(Y, q)$ which results in
   $$15q + 8(1-q) = 8q + 13(1-q),$$
   that is $q = 5/12$;
   - for player J, the condition is $u_K(p, M) = u_K(p, 0)$ which results in
   $$5p + 8(1-p) = 8p + 3(1-p),$$
   that is $p = 5/8$.

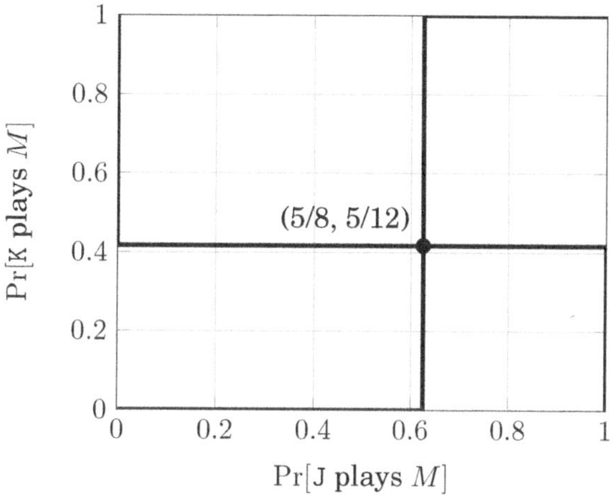

Figure 1.2: Mixed Nash equilibrium of the game.

Exercise 1.2. Mr. Square (S) and Ms. Triangle (T) choose one point in an xy-graph. Their choices are made independently and unbeknownst to each other, but they are fully aware of each other's rules of choice. After their choices are made, they get the following rewards computed as revenue minus cost:

- if they chose the same point, they both get no revenue; if they chose a different point, they both get a revenue equal to 10
- the cost is computed as $x + 3y$ where x and y are the coordinates of the chosen point

Player S can choose one of these four points: A $(0, 0)$; B $(0, 4)$; C $(4, 4)$; D $(4, 0)$.
Player T can choose one of these three points: A $(0, 0)$; E $(2, 4)$; D $(4, 0)$.

Answer the following questions:

1. What kind of game is this? Write down its normal-form representation.
2. Find out all the Nash equilibria in pure strategies of this resulting game.
3. Find out all the Nash equilibria of this resulting game.

Solution

1. This is a static game of complete information. The normal-form representation can be shown as a bi-matrix, where the set of players is {S, T}. Their strategy sets are $\mathcal{S}_S = \{A, B, C, D\}$ and $\mathcal{S}_T = \{A, D, E\}$.

S \ T	A	D	E
A	0, 0	10, 6	10, -4
B	-2, 10	-2, 6	-2, -4
C	-6, 10	-6, 6	-6, -4
D	6, 10	-4, -4	6, -4

2. It is immediate to find that B and C (for player S) and E (for player T) are dominated strategy (e.g., by A). So, players can only play A or D and it turns out by inspection that (A, D) and (D, A) are indeed Nash equilibria in pure strategies.

S \ T	A	D
A	0, 0	10, 6
D	6, 10	-4, -4

3. There is one mixed-strategy Nash equilibrium equilibrium: it is easy to spot it, because the game is similar to a coordination game like the Battle of Sexes, but with the goal of choosing a different strategy for both (i.e.,

not to meet in the same point). This is actually called an *anti-coordination game*. To find the mixed strategy equilibrium, just set α as the probability of choosing strategy A, and thus D is played with probability $1 - \alpha$. For symmetry reasons, α is the same for both players. Finding it can be done through the indifference principle stating that $u_S(A, \alpha) = u_S(D, \alpha)$, which results in $10 - 10\alpha = 6\alpha - 4 + 4\alpha$

$$\alpha \cdot 0 + (1 - \alpha)10 = \alpha \cdot 6 + (1 - \alpha) \cdot (-4),$$

and thus $\alpha = 0.7$.

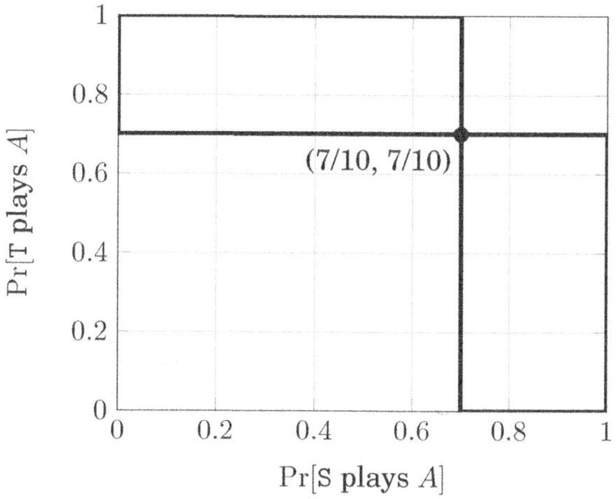

Figure 1.3: Mixed Nash equilibrium of the game.

Exercise 1.3. A radio transmitter-receiver pair works without coordination for a single communication. The transmitter can either: concentrate its transmission on a single frequency f_0 (C) or spread its transmission over the entire spectrum (S). Just using f_0 costs 2 units of power, while using the spread spectrum costs 9 units of power. The receiver can listen to one frequency f_1 (A), or two frequencies f_1 and f_2 (B). Note that it is not known whether $f_0 = f_1$ or $f_0 = f_2$. If the transmitter uses S, the utility of the receiver is 2 plus the number of frequencies used. If the transmitter uses C, the utility of the receiver is 2 minus the number of frequencies used. The utility of the transmitter is equal to twice the utility of the receiver plus the number of frequency used by the receiver and minus the power cost. Also, note that the pure strategies described above (for both players) can be meaningfully combined into a mixed strategy.

1. Draw the normal form of the game.
2. Find all pure strategy Nash equilbria of the game.
3. Does the game have a mixed strategy Nash equilibrium? If yes, find it. If not, prove why.

Solution

1. The set of players is {T, R}, with respective strategy sets $\mathcal{S}_T = \{C, S\}$ and $\mathcal{S}_R = \{A, B\}$. The payoffs of the normal form can be represented in a bi-matrix as follows.

R

T		A	B
	C	1, 1	0, 0
	S	-2, 3	1, 4

2. Nash equilibria are found when the strategy played by each player is a best response to the other. A's best responses to B's move are underlined, B's best responses are overlined.

R

T		A	B
	C	$\overline{\underline{1, 1}}$	0, 0
	S	-2, 3	$\overline{\underline{1, 4}}$

This is a coordination game and has two pure strategy Nash equilibria: (A,C) and (B,S).

3. There is a mixed strategy Nash equilibrium. It can be found by setting p as the probability that the transmitter plays C and q as the probability that the receiver plays A.
Thus, for T: playing C gives utility q and playing S gives utility $-2q + 1 - q$.

C is preferable if $q > 1 - 3q$, that is, $q > \frac{1}{4}$.
For R: playing A gives utility $p + 3(1-p)$ and playing B gives utility $4(1-p)$. A is preferable if $3 - 2p > 4 - 4p$, that is $p > \frac{1}{2}$. The mixed Nash equilibrium is found as $p = \frac{1}{2}$, $q = \frac{1}{4}$.

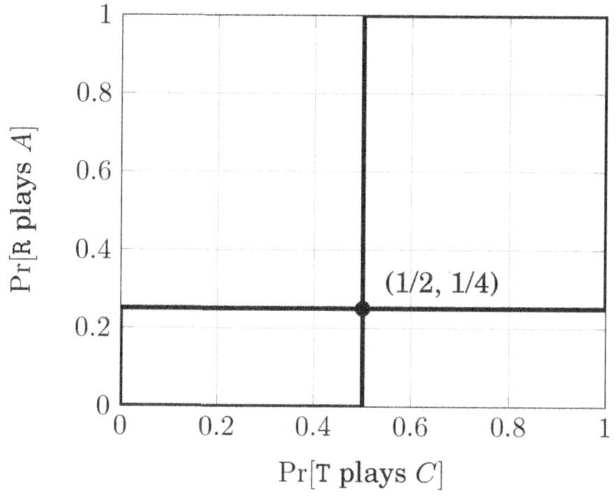

Figure 1.4: Mixed Nash equilibrium of the game.

Exercise 1.4. Two students, Charlotte and Daniel, need to write their MS Thesis. They need to choose (independently and unbeknownst to each other) a supervising professor. Three professors are available for this role: Xavier, Yuan, and Zingberry. The utility of a student is given by the amount of help he receives from his supervisor, which is quantified as 40 for Xavier, 60 for Yuan, 50 for Zingberry. However, if the two students select the *same* professor as their supervisor, they only get 70% of the utility that they would get if the professor had only one of them to supervise.

1. Write the game in normal form.
2. How many Nash equilibria does this game have?
3. If your answer of the previous point included one or more mixed equilibria, write all of them explicitly (i.e., evaluate the probabilities that each professor is chosen).

Solution

1. The set of players is {C, D}, and the common set of strategies is $\mathcal{S}_C = \mathcal{S}_D = \{X, Y, Z\}$. The normal form of the game can be represented as a bi-matrix as follows.

C \ D	X	Y	Z
X	28, 28	40, 60	40, 50
Y	60, 40	42, 42	60, 50
Z	50, 40	50, 60	35, 35

2. Strategy X is strictly dominated by Y. Thus, we can remove it. We are left with the following reduced game:

C \ D	Y	Z
Y	42, 42	60, 50
Z	50, 60	35, 35

   This game has 2 Nash equilibria in pure strategies: (Y,Z) and (Z,Y).

3. It is easy to see that there must also be a third NE in mixed strategies, where the support of both mixed strategies includes Y and Z, while X is discarded as it is strictly dominated. Since the matrix is simmetric, the mixed strategies that are played in the Nash equilibrium must be equal for both the players. If p is the probability of either of the players choosing Y, the mixed strategy equilibrium must satisfy the following condition:

$$42p + 60(1 - p) = 50p + 35(1 - p)$$

which leads to $33p = 25$ and hence $p = \frac{25}{33}$. Thus, the mixed Nash equilibrium consists in both players adopting the strategy "play X with probability 0, Y with probability p, Z with probability $1 - p$", where $p = \frac{25}{33}$.

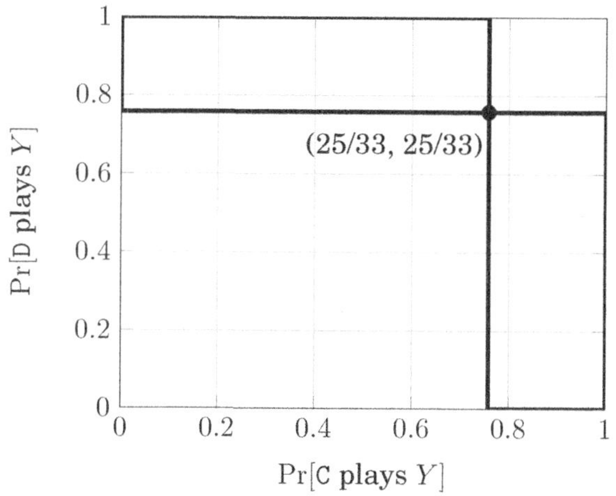

Figure 1.5: Mixed Nash equilibrium of the game.

Exercise 1.5. Alice (A) and Bob (B) want to play together War Fantasy World, a very popular online game. When players create a character for the game, they are asked to select a race, and 4 options are available: Dwarves (D), Elves (E), Humans (H), and Orcs (O). Alice and Bob create characters simultaneously and independently. Their combined choice gives them utilities, whose values are the benefit of the teamplay, minus the cost. The cost is 0 for Humans, 3 for Orcs, 4 for Dwarves, 5 for Elves.
Humans are vanilla characters; they do not give a benefit to others.
Orcs give to the other player a benefit of -3 (i.e., decrease their benefit by 3).
Dwarves give an additional benefit of $+5$ to the other player.
Elves give a benefit of $+15$ to the other player only if they are an Elf too.
Finally, Alice gets an extra benefit of $+5$ if she plays an Orc character but Bob is not playing as an Orc too. All of these rules are common knowledge.

1. Write down the normal form of the game.
2. Find the Nash equilibria of this game in pure strategies.
3. Find any additional Nash equilibria of this game in mixed strategies.

Solution

1. The set of players is $\{A, B\}$, with common strategy set $\mathcal{S}_A = \mathcal{S}_B = \{H, E, D, O\}$. Payoffs are:

B

A	H	E	D	O
H	0, 0	0, -5	5, -4	-3, -3
E	-5, 0	10, 10	0, -4	-8, -3
D	-4, 5	-4, 0	1, 1	-7, 2
O	2, -3	2, -8	7, -7	-6, -6

By applying IESDS, strategy D can be eliminated as it is strictly dominated by H for both players. Also strategy O can be eliminated for player B. Thus, strategy H is eliminated for player A. This reduced matrix survives:

B

A	H	E
E	-5, 0	10, 10
O	2, -3	2, -8

2. The Nash equilibria are (O, H) and (E, E).
3. The mixed-strategy equilibrium is found with Alice playing E with probability p and O with $1 - p$; Bob playing H with probability q and E with $1 - q$. Applying the indifference principle Alice's payoff at the equilibrium satisfies

$$-5q + 10(1 - q) = 2$$

so $q = 8/15$, and Bob's payoff satisfies

$$-3(1-p) = 10p - 8(1-p)$$

so $p = 1/3$.

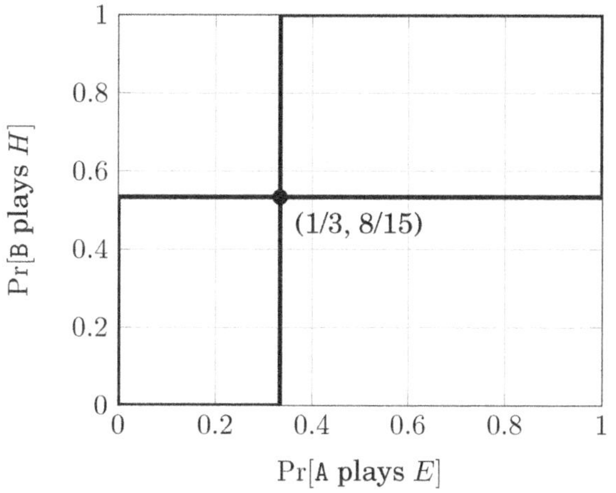

Figure 1.6: Mixed Nash equilibrium of the game.

Exercise 1.6. Two sport teams (T1 and T2) play against each other in an exhibition match. Each team can choose a style of playing: aggressive (A); spectacular (S); or relaxed (R). Important: regardless of the actual number of components in a team, both teams are considered as single players in the analysis, since they decide a style for the match (and this decision applies to the whole team), and they do so independent and unbeknownst to each other. At the end of the match, each team gets a payoff based on both styles chosen. Also note that this payoff does not necessarily represent the score in the match (which is just an exhibition) but rather whether they enjoyed the match, had a good time, and so on. A team playing aggressively always gets a basic payoff equal to 1. A team playing in a spectacular way always gets payoff equal to 3 and in addition increases the payoff of the other team by 1. A team playing in a relaxed way gets payoff 0, unless *both* teams play relaxed, in which case they both get payoff 5.

1. Draw the normal form of the game.
2. Find all pure strategy Nash equilibria of the game.
3. Does the game have a mixed strategy Nash equilibrium? If yes, find it. If not, prove why.

Solution

1. The set of players is {T1, T2} with common set of strategies $\{A, S, R\}$. The normal form is fully characterized by a bi-matrix as follows.

T2

T1	A	S	R
A	1, 1	2, 3	1, 0
S	3, 2	4, 4	3, 1
R	0, 1	1, 3	5, 5

2. Pure strategy A is strictly dominated by S, so we can safely remove it.

T2

T1	S	R
S	4, 4	3, 1
R	1, 3	5, 5

The game is left with two pure strategy Nash equilibria: (S,S) and (R,R).

3. There is a mixed strategy Nash equilibrium. It can be found by setting p as the probability that team T1 plays S.
Then, for team T2, playing S gives utility $4p + 3(1 - p) = p + 3$ and playing R gives $p + 5(1 - p) = 5 - 4p$.

S is preferable if $p + 3 > 5 - 4p$, which means $p > \frac{2}{5}$. The reasoning can be repeated for reversed roles of teams T1 and T2. The mixed strategy Nash equilibrium is therefore that each team plays S with probability $\frac{2}{5}$ (and R with probability $\frac{3}{5}$).

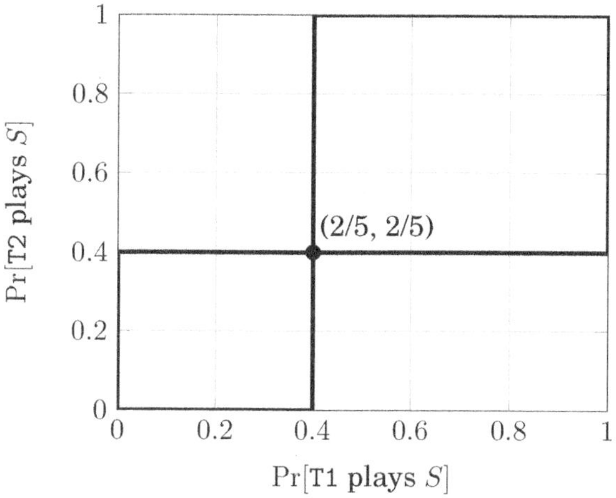

Figure 1.7: Mixed Nash equilibrium of the game.

Exercise 1.7. Two firms (F1 and F2) work on a joint project from the European Commission. They can allocate an integer number of employees on the project, from 0 to infinity. They decide independently and without consulting with each other. The outcome of the project is that, if the number of employees allocated by each firm is identical (even zero!), both firms get a funding of 140 k€ from the European Commission. If the two firms assign a different number of employees, the European Commission gives them different fundings: 350 k€ to the one with more employees, and 160 k€ to the one with fewer employees. However, assigning employees costs 100 k€ per employee. The *utility* of a firm is funding minus costs.

1. Show that no rational firm will allocate more than 2 employees on the project.
2. Draw the normal form of this game (considering then up to 2 employees) and find its Nash equilibria in pure strategies.
3. Find the additional Nash equilibria in mixed strategies.

Solution

1. By playing 0, a firm gets a security payoff of 140 k€. The best one can get by assigning 3 employees is 50 k€. The same reasoning can be done for any other number greater than 2, so a rational firm will never allocate more than 2 employees.
2. The set of players is {F1, F2}. The strategy sets are $S_1 = S_2 = \{0, 1, 2\}$, because we can safely discard 3 or more, as per the previous discussion. The payoffs are:

		F2: 0	F2: 1	F2: 2
F1	0	140, 140	160, 250	160, 150
	1	250, 160	40, 40	60, 150
	2	150, 160	150, 60	-60, -60

The Nash equilibria are found to be $(1, 0)$ and $(0, 1)$. Indeed, 0 is the best response to anything but 0 and the best response to 0 is 1. We can also see that 2 is not strictly dominated by 0 and 1 but it is dominated, e.g., by their convex combination where 0 is played with probability $p = 0.95$ and 1 is played with probability $1 - p = 0.05$. Thus, we can also remove 2 from the game.

		F2: 0	F2: 1
F1	0	140, 140	160, 250
	1	250, 160	40, 40

3. At this point, there is only one Nash equilibrium left in mixed strategy. It can be found by setting the strategy of firm 1 to be to play 0 with probability p and 1 with probability $1-p$. The payoff of F2 is then $140p+160(1-p)$ when playing 0 and $250p+40(1-p)$ when playing 1 and they must be equal. So $p = 12/23$.

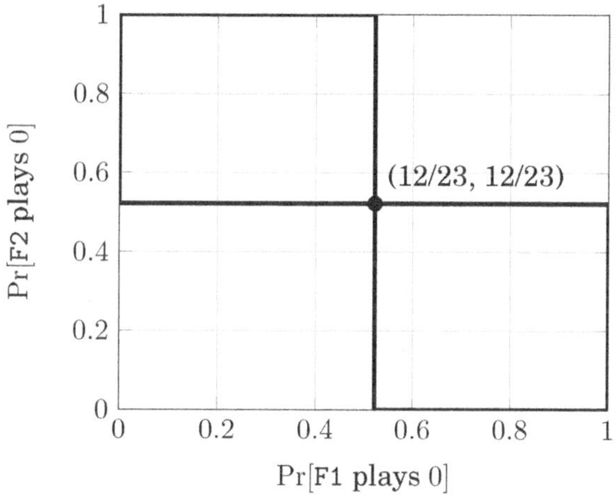

Figure 1.8: Mixed Nash equilibrium of the game.

Exercise 1.8. Andrea (A) and Brianna (B) are planning a two-week trip in Italy together. They agreed that A and B will book a hotel for the first and second week, respectively. They will also make their decision unbeknownst to each other (when they find out the other's choice, it is too late to get a refund or change booking). There are available hotels in: Florence (F), Rome (R), or Venice (V). Both girls would like to visit Tuscany (to see Pisa, Lucca, Siena). However, this is possible only if they manage to book one week in Florence and one week in Rome (not necessarily in this order). If they both book F or they both book R, they cannot visit Tuscany; same if either of them books V.

Final payoffs are computed as follows. If they manage to visit Tuscany, both get utility 3. If they just visit Florence, but not the remainder of Tuscany, A gets utility 1 and B gets 0. If they just visit Rome instead, the situation is reversed: A gets nothing, and B gets 1. Finally, for every week they spend in Venice, increase both utilities by 1. That is, if their choices are (F,V), their final utilities are 2 for Andrea and 1 for Brianna. If they spend two weeks in Venice, both get utility 2.

1. Formalize this as a static game and represent it in normal form.
2. Find out all the Nash equilibria of this game in pure strategies.
3. Find *one* of the possible mixed strategies Nash equilibria.

Solution

1. The set of players is {A, B}. The common strategy set is $\mathcal{S}_A = \mathcal{S}_B = \{F, R, V\}$. Payoffs are represented through the following matrix:

A \ B	F	R	V
F	1, 0	3, 3	2, 1
R	3, 3	0, 1	1, 2
V	2,1	1, 2	2, 2

2. The game has 9 possible outcomes in pure strategies. The Nash equilibria are (F, R), (R, F) and (V, V).

3. The game has several mixed strategies Nash equilibria. For example, you can consider a mixed strategy where both players play all of the pure strategies F, R, V with non-zero probability.

   If we call f and r the probabilities that Andrea plays F and R, respectively (then the probability of playing V is $1 - f - r$), the condition to impose is

$$\begin{cases} u_B((f, r, 1-f-r), F) = u_B((f, r, 1-f-r), R) \\ u_B((f, r, 1-f-r), V) = u_B((f, r, 1-f-r), R) \end{cases}$$

   which leads to

$$\begin{cases} 0f + 3r + 1 - f - r & = 3f + r + 2(1-f-r) \\ f + 2r + 2(1-f-r) & = 3f + r + 2(1-f-r) \end{cases}$$

with solution $f = 1/4$ and $r = 1/2$. Applying the same reasoning for Brianna, imposing $u_A(F, (\phi, \rho, 1 - f - \rho)) = u_A(R, (\phi, \rho, 1 - f - \rho)) = u_A(V, (\phi, \rho, 1 - f - \rho))$, we get $\phi = 0.5$ and $\rho = 0.25$.

Other possible equilibria can be found by ignoring strategy V, and setting its probability to 0. This leads to a simpler system, which is basically a discoordination game, since the players get the most when they pick different strategies. The solution here is obtained by imposing $u_B((f, 1 - f, 0), F) = u_B((f, 1 - f, 0), R)$ and $u_A(F, (\phi, 1 - \phi, 0)) = u_A(R, (\phi, 1 - \phi, 0))$, so we get $f = 0.4$ and $\phi = 0.6$.

Exercise 1.9. Atlas network (A) and Broadcasting Ltd (B) are local TV channels, competing for the most audience. They both have three news broadcasts in the morning (M), daytime (D), and evening (E). They must decide what is the time share where they invest the most, and can only choose one – without of course telling each other. If they choose the same slot, the payoff is 0 for both. If they choose a different one, there is a "winner", i.e., whoever chooses the earlier slot. For example, between M and D, the winner is whoever chooses M. However, E beats M because the evening is actually earlier than the following day's morning and it allows to give the news earlier than the competitor. The losing channel always gets -1. The winning channel gets a reward depending on how valuable is that time slot to advertising, so it is $+1$, $+4$, $+2$ if the winning move is M, D, E, respectively.

1. Write down this game in normal form.
2. Find the Nash equilibria in pure strategies.
3. Find a Nash Equilibrium in mixed strategies.

Solution

1. The game is basically a modified rock-paper-scissors. Differently from that game, it is *not* zero-sum (see the next chapter), but it is is competitive, in that whenever the payoff of a TV channel is increased, the other decreases. The set of players is {A, B}, and the strategy sets are $\mathcal{S}_A = \mathcal{S}_B = \{M, D, E\}$. The payoffs are represented in the bi-matrix below.

B

A		M	D	E
	M	0, 0	1, -1	-1, 2
	D	-1, 1	0, 0	4, -1
	E	2, -1	-1, 4	0, 0

2. As in rock-paper-scissors, there is no Nash equilibrium in pure strategies (there will always be at least one player regretting their choice).

3. Because of Nash theorem, a Nash equilibrium must exist. It can be found by setting probabilities of playing M, D, E as m, d, $1 - m - d$, respectively. Because of symmetry, these must be the same probablities for both players. Applying the indifference principle, we can write the equations

$$\begin{cases} d - 1 + m + d = -m + 4 - 4m - 4d \\ d - 1 + m + d = 2m - d \end{cases}, \tag{1.1}$$

with solutions $\mu = 9/24$ and $\delta = 11/24$. The mixed-strategy NE consists then in A and B playing M, D, E with probabilities $9/24$, $11/24$, and $4/24$.

Exercise 1.10. A 90-year old grandmother writes on the family messaging chat: "Oh! It looks like I am out of eggs to prepare my special cake. Can somebody go to the grocery store and fetch me some?" The only family members that can answer this request are grandchildren Abel, Bella, and Claire (A, B, and C). They know that they are the only three people that can help grandma, and everything else said below is also common knowledge to them. They need to decide on the spot whether to get eggs at the grocery store for grandma (G) or not (N), without consulting each other. They all are aware that going to the grocery store has a cost $c = 1$. And grandma does not need all of them to buy eggs, even if only one goes, she will be able to prepare the cake anyways. However, if nobody fetches the eggs, grandma is unable to prepare her special cake, hence payoff 0 for the entire family. If grandma prepares the cake instead, the payoff is 4 for everybody (including those who did not help her). The best option for A, B, and C is of course that somebody else helps grandma and they just eat the cake.

1. Write down this game in normal form.
2. Find the Nash equilibria in pure strategies.
3. Find a Nash equilibrium in mixed strategies where all players may play both actions with non-zero probabilities.

Solution

1. The set of players is $\{\mathtt{A}, \mathtt{B}, \mathtt{C}\}$. The strategy set is $S = \{G, N\}$ for all of them. The payoff matrix is in three dimensions (a $2\times2\times2$ cube) and can be written as follows:

Case A plays G

B \ C	G	N
G	3, 3, 3	3, 3, 4
N	3, 4, 3	3, 4, 4

Case A plays N

B \ C	G	N
G	4, 3, 3	4, 3, 4
N	4, 4, 3	0, 0, 0

2. The Nash equilibria in pure strategies are (G, N, N), (N, G, N), (N, N, G). Indeed, the only cases in which nobody has regrets are when just one of the family members goes to the grocery and the others do not.

3. To find the Nash equilibrium in mixed strategies, we assume that A plays G with probability p (and N with probability $1 - p$) and so do B and C, for symmetry reasons. The indifference principle guarantees that each player must be indifferent between playing G that guarantees payoff 15 as discussed above, and N. What is the expected payoff when playing N? Assume, without loss of generality, that A plays N. With probability $(1-p)^2$ neither B nor C go fetch the eggs as well, so the payoff is 0. With probability $1 - (1 - p)^2 = 2p - p^2$ somebody else than A fetches the eggs, in which case

A's payoff would be 4. Thus, we must require $3 = 8p - 4p^2$, which implies $p^2 - 2p + 0.75 = 0$. The solution is $p = 1 - \sqrt{0.25} = 0.5$. There is also a solution $p = 1.5$ but it is not acceptable as p, being a probability, must be lower than 1. So the resulting NE in mixed strategies is that all players choose the same strategy, that is, with probability $p = 0.5$ they play G and with probability $1 - p = 0.5$ they play N.

Exercise 1.11. Agatha (A) and Bruno (B) are two siblings. They own a model train set (T) and they usually play together with it. Their respective utilities when doing so is $u_{\mathtt{A}}(T,T) = u_{\mathtt{B}}(T,T) = 3$. Their mom (M) buys them a new toy (N), which can be either a dollhouse (D) or a set of small soldier figurines (S). The two kids must independently decide whether to keep playing with T or the new toy N. All the three players (A, B, M) involved decide their action simultaneously and unbeknownst to each other. The action set of the kids includes just T or N; the mom instead decides what new toy to buy. If the kids end up in not playing together, *all* the players (also M) get utility equal to 0. If they play together with the new toy, they get a positive utility that for A is equal to 6 and 1 for D and S, respectively; for B, the values are instead 1 and 4, respectively. The utility of M is the lump sum of the utility of the two kids.

1. Write down the normal form of this game (likely, you will need a 3D matrix: you can write two matrices instead, one per each move of player M).
2. Find all Nash equilibria of this game in pure strategies.
3. Find all additional Nash equilibria of this game in mixed strategies.

Solution

1. The set of players is $\{\mathtt{A},\mathtt{B},\mathtt{M}\}$. Their strategy sets are $\mathcal{S}_{\mathtt{A}} = \mathcal{S}_{\mathtt{B}} = \{T,N\}$, and $\mathcal{S}_{\mathtt{M}} = \{D,S\}$. Payoffs can be represented using a matrix in three dimensions as follows:

Case M plays D

A \ B	N	T
N	6, 1, 7	0, 0, 0
T	0, 0, 0	3, 3, 6

Case M plays S

A \ B	N	T
N	1, 4, 5	0, 0, 0
T	0, 0, 0	3, 3, 6

2. The game has 8 possible outcomes in pure strategies. By inspection, it is clear that none of the solutions where the kids do not coordinate can be a Nash equilibrium. This leaves open four cases: of these, $(T,T,\cdot)$ are Nash equilibria, since no player has an incentive to unilaterally deviate. Also this is true for (N,N,D) but not for (N,N,S), since M has an incentive to deviate and play D instead.

3. First of all, note that technically speaking, if both kids choose T, M choice is arbitrary and can even be a mixed strategy. So there are infinitely many mixed strategy NE of the form $(T,T,\cdot)$, where the last move can be any combination, they all will obtain the same payoff. Furthermore, observe that D is weakly dominant for M, where the "weakness" of the inequality is only given because the two available actions of M get the same payoff only in the case both kids play T, as per the discussion above. In any other non-degenerate linear combination of A and B's choices, M is always strictly

better off by playing D. Thus, we discard S and consider just the left-hand matrix above. This is now a coordination game (analogous to the Battle of sexes). To find a mixed equilibrium (p, q) (p and q being the respective probabilities of A and B playing N) the characterization property requires that $u_{\mathtt{A}}(N, q) = u_{\mathtt{A}}(T, q)$, $u_{\mathtt{B}}(p, N) = u_{\mathtt{B}}(p, T)$, that leads to

$$\begin{aligned} 6q &= 3(1-q), \\ p &= 3(1-p), \end{aligned}$$

i.e., $p = 3/4$ and $q = 1/3$.

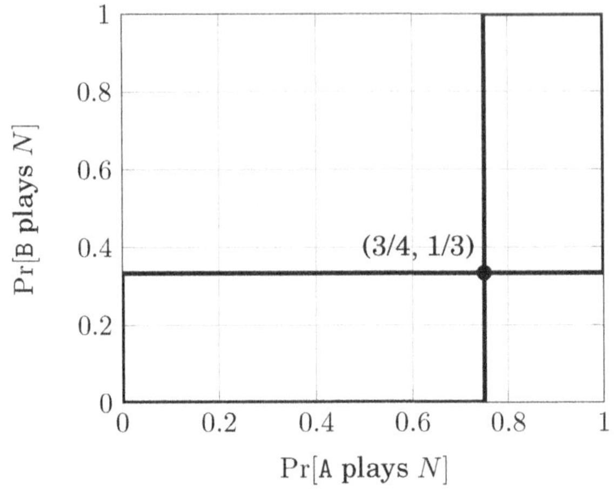

Figure 1.9: Mixed Nash equilibrium of the game.

Thus, the resulting additional mixed strategy NEs are: $(T, T, \cdot)$, with M playing any mixture of her strategies, and $(\frac{3}{4}N + \frac{1}{4}T, \frac{1}{3}N + \frac{2}{3}T, D)$

2 Zero-sum games and minimax

Zero-sum games represent strategic interactions between two players, where their payoffs sum to 0 for every possible outcome. They allow for specific solution approaches that may be more convenient than the general methods. The concepts of maximin and minimax, which are particularly poignant for this class of games, can also be generalized and associated to other scenarios. Overall, zero-sum games and maximin/minimax are also typically encountered as applications of linear programming and artificial intelligence.

2.1 Characteristics

Zero-sum games are a specific case of competitive strategic interaction, for which some notable theorems hold. This guarantees better computational properties to the problem of finding the NE that is generally hard.

The formal definition of zero-sum, as the name suggests, imples that, whatever the final outcome of the game, the sum of the payoffs of the two players will always be zero. This is true for all situations where two players compete, and whatever one wins, it is lost by the other. Games like "Odds and Evens," chess, or poker are therefore zero-sum (but note that, while sharing this property, they are quite different in many other aspects). In some engineering applications (but also when dealing with pre-school kids), non-zero-sum cases may instead be appealing if they include a possible outcome where everyone can win.

Several theoretical results can be proven for zero-sum games, most notably the minimax theorem, which puts them in close relationship to linear programming duality and implies that polynomial-time procedures for finding the NE exist, which is not true for a generic game. In general, a zero-sum game can also be seen in many engineering problems with an adversarial setup, such as those related to network security. Here, some network performance metric is quantified and one player, representing the legitimate network user, tries to maximize it, hence it is called the *maximizer*, whereas an attacker plays the role of the *minimizer*, i.e., someone who wants to decrease the performance metric to its lowest possible values.

Zero-sum games are actually a special case of a more general class that shares the same descriptive principles of the players being in competition against each other, but whose sum of the payoff is not actually 0. These are sometimes called competitive or conflict games. Proper utility transformations can leave the NE unaltered but change such setups into proper zero-sum games.

For example, consider a game where two players compete over a resource R to be shared among them, like a delicious pie. Whatever share x is obtained by player 1, player 2 will get $R-x$. So technically speaking the game is R-sum rather than zero sum, but the utilities can be rescaled (e.g., considering the difference from the value $R/2$) to obtain a zero sum game.

Or, one can also consider the old rules for assigning points in football leagues after individual matches. Until the 80s, two points where usually assigned per match, such that a winner got 2 points, a loser got nothing, and in case of a tie, the two players “split the pot”. Hence, this is the same situation as above with $R = 2$. To increase competition, this rule was changed to assigning 3 points for a victory, and the game can no longer be translated into a proper zero-sum setup, even though its competitive character is clear. In practice, competitive games are situations where all outcomes are Pareto efficient, since a player cannot improve the payoff without making the other player worse.

2.2 Maximin and minimax

If player i faces other players generically described as $-i$ (as typical of game theory), and the strategies of these players are described by s_i and s_{-i}, respectively, and finally $u_i(s_i, s_{-i})$ is the utility of player i, the maximin and the minimax for player i are computed as

$$\text{maximin}_i = \max_{s_i} \min_{s_{-i}} u_i(s_i, s_{-i}) \qquad \text{minimax}_i = \min_{s_{-i}} \max_{s_i} u_i(s_i, s_{-i}), \tag{2.1}$$

where the reason for their names is clear from the definition.

The maximin (also called the security payoff) represents the maximum payoff that i can achieve even under conservative (i.e., worst-case) assumptions, without knowing the strategies chosen by the other players. The minimax instead is the guaranteed payoff that player i can get if knowing the strategies of $-i$, since i is a rational player that will therefore choose the best response, and the minimax is therefore the lowest of the payoff obtained by i when playing the best response.

These concepts are relevant in the context of the minimax theorem, which combines two properties. The first is that, whenever the minimax and the maximin coincide *for a zero-sum game*, their value also corresponds to the NE payoff (the NE is not necessarily unique, but all NE get the same payoff). Note that this result is in general not true for games that are not zero-sum. Moreover, if the strategies s_i and s_{-i} are chosen as mixed strategies, and thus u_i is computed as an expected utility, the continuity of the linear combination within the expected value necessarily makes the maximin and the minimax coincide when computed over mixed strategies, which is an alternative way to see the Nash theorem for the existence of an NE, however, with the additional property that all the NEs share the same payoff value, which is a consequence of the game being zero-sum.

Moreover, a linear programming approach exploiting this theorem leads to a computationally efficient derivation of the NE for zero-sum games. This corresponds to considering a linear combination of the utilities at the pure strategies of the two players. If the maximizing and the minimizing players have L and M available pure strategies, respectively, the optimization variables are the weights of the L pure strategies of the maximizer, and M constraints can be considered for each of them, where the linear combination is greater than or equal to a slack variable w. In turn, the slack variable w is the maximization goal. Note that this approach obtains at the same time the value of the payoff at the NE in mixed strategies and the support of the mixed strategy NE itself, which is identified by the constraints that are hit when w is maximized.

Finally, a dual approach is also possible, where the roles of the maximizer and the minimizer are reversed (and so the objective of the optimization is to minimize rather than to maximize), which, differently from before, achieves instead a problem with M variables and L constraints. The ultimate choice of what approach is more convenient depends on the numerical values of L and M.

Exercises

Exercise 2.1. Coral and Dwayne are playing a game where they both have three cards in hand, showing a number, i.e. the *face value* of the card; they must select one and play it. Each player knows what cards do the opponent have in hand (as well as theirs). After the cards are drawn, players compute their points as follows.

If they select two cards with the same face value F, Coral gets F points and Dwayne gets $-F$ points. If they select cards with different face values respectively equal to F_C and F_D, Coral gets $F_C - F_D$ and Dwayne gets $F_D - F_C$ points.

Coral cards are a Nine (9), a Six (6) and a Two (2). Dwayne cards are a Nine (9), an Eight (8) and a Two (2). This information is common knowledge, as well as the rules of the game.

1. Formalize this as a static game and represent it in normal form (if there are simplifications of the bi-matrix available, use them in the representation).
2. Find out all the Nash equilibria of this game in pure and mixed strategies.
3. Find the minimax and the maximin for both players.

Solution

1. The set of players is $\{\mathtt{C}, \mathtt{D}\}$. Their strategy sets are $\mathcal{S}_\mathtt{A} = \{9, 6, 2\}$ and $\mathcal{S}_\mathtt{B} = \{9, 8, 2\}$.

   The game is zero-sum, Payoffs are represented through the following matrix:

   D

C		9	8	2
	9	9, -9	1, -1	7, -7
	6	-3, 3	-2, 2	4, -4
	2	-7, 7	-6, 6	2, -2

   However, the game is zero-sum, so a better representation just shows Coral's payoffs (u_C), and Dwayne's payoffs are derived as $-u_C$.

   D

C		9	8	2
	9	9	1	7
	6	-3	-2	4
	2	-7	-6	2

2. It is immediate to see that 2 is a dominated strategy for both players. For Coral, playing 9 is dominant. Thus, the only Nash equilibrium (which is in pure strategies) is (9,8).

3. Since the payoff at the Nash equilibrium is 1 for Coral, and the game is zero-sum, it must be that $\text{minimax}_C = \text{maximin}_C = 1$, and $\text{minimax}_D = \text{maximin}_D = -1$. This can also be verified directly. In order to find the maximin of C, one must observe the rows of the matrix (looking at C's payoffs) finding the minimum value for each row, and choose the maximum among those values.

C \ D	9	8	2	
9	9	1	7	→ 1
6	-3	-2	4	→ -3
2	-7	-6	2	→ -7

⇒ Maximin of C: 1

In order to find the maximin of D, one must observe the columns of the matrix (looking at D's payoffs) and apply the same procedure.

C \ D	9	8	2
9	-9	-1	-7
6	3	2	4
2	7	6	-2
	↓ -9	↓ -1	↓ -7

⇒ Maximin of D: -1

The procedure to find the minimax is the dual of the maximin: one must look at columns instead of rows for C (and vice versa for D), finding first the maxima and then the minimum among those.

Exercise 2.2. Consider the following game in normal form. Players are denoted as 1 and 2 and their strategy sets are $\mathcal{S}_1 = \{A, B, C\}$ and $\mathcal{S}_2 = \{X, Y\}$. The payoff matrix is as follows:

2

1		X	Y
	A	9, -9	-5, 5
	B	-2, 2	7, -7
	C	8, -8	-1, 1

1. Describe what kind of game is that and name a specific solution concept you might use to find its Nash equilibria.
2. What is the support within $\mathcal{S}_1$ of the mixed strategy played at the Nash equilibrium by 1?
3. Find the mixed strategies played by 1 and 2 at the Nash equilibrium.

Solution

1. The game is zero-sum and can be solved through minimax search. Indeed, it must have a Nash equilibrium in mixed strategies (only one, since 2 has only one way of mixing its strategy). Specifically, let be b the probability of player 2 to play X. Then, for 1 the minimax conditions are:

$$\begin{cases} 9b - 5(1-b) \geq w \\ -2b + 7(1-b) \geq w \\ 8b - 1(1-b) \geq w \end{cases}$$

where w is a slack variable to be maximized.

2. A graphical inspection reveals that the active constraints are the 2nd and 3rd, which means that the support of the mixed equilibrium is (B, C) for player 1.

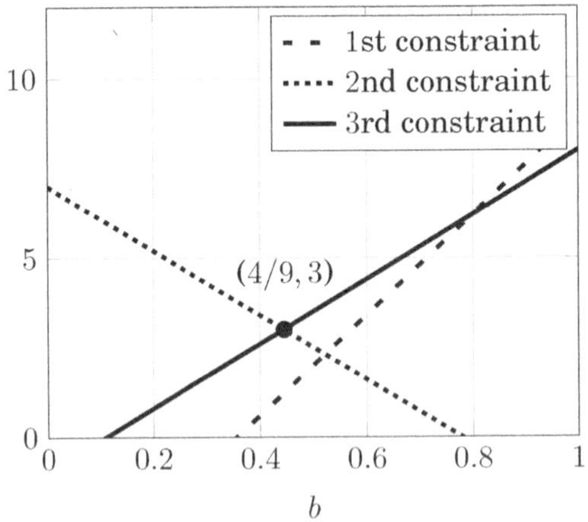

Figure 2.1: Visual representation of the constraint boundaries. Maximizing w with respect to the last two constraints satisfies also the first, hence the Nash equilibrium is the interception of the two lines.

3. We can then equate $7-9b = 9b-1$ which results in $b = 4/9$. The payoff at the Nash equilibrium for 2 is -3. Consequently, denoting with a the probability of Player 1 choosing B within the support, we have: $a \cdot 3 = (1-a) \cdot 3$, i.e., $a = 1/2$. The Nash equilibrium is $(0, 1/2, 1/2); (4/9, 5/9)$ and the value is 3.

Exercise 2.3. A tournament match is about to be played by the Azure (A) team versus the Brown (B) team. The teams have more or less the same strength, so the winner will only be determined by the strategical choices of the coaches; these choices are to be made before the match starts and without the other team knowing it. Both coaches have three options for the way their teams will be playing in the upcoming match: counter-offensive (C), defensive (D), or extremely aggressive (E). If they both play the same choice, the game will end in a tie (zero points for both teams). A team choosing to play counter-offensive will beat a team playing extremely aggressive, in which case the winner gets three (3) points. A defensive team will defeat a team playing counter-offensive, getting one (1) point. Finally, an extremely aggressive team will win over a defensive team, and will get eight (8) points. The loser always get a negative amount of points, whose absolute value is the same as the winner's.

1. Write the normal-form representation of this game. You can probably exploit some special structure of the game to simplify the matrices involved; explain it if you do so.

2. Find the minimax of this game in mixed strategies and justify why you get this value.

3. Find the Nash equilibria of this game in both pure and mixed strategies, by exploiting the results previously found.

Solution

1. The set of players is {A, B}, with common set of strategies $\{C, D, E\}$. The game is zero-sum by construction. Thus, only A's payoffs can be written (B's payoffs are minus that). The resulting matrix is:

B

A	C	D	E
C	0	-1	3
D	1	0	-8
E	-3	8	0

2. The minimax in mixed strategies is zero. A simple answer to that can be because the game is symmetric. A more detailed answer: since the three strategies are in a Condorcet cycle (C beats E that beats D that beats C) there cannot be a pure-strategy min-maximizer and not even a min-maximizer where a player plays only a mixture of two pure strategies. Hence, the minimax must happen for a mixed strategy that is a convex combination of all three strategies C,D,E.

3. In pure strategies, the maximin is -1 and the minimax is 1. The minimax in mixed strategies is zero, and the strategy at Nash equilibria can be denoted

by (c, d, e) for both players, as a triple of playing C, D, E, respectively. The numerical values are found by solving:

$$\begin{cases} d - 3e = 0 \\ -c + 8e = 0 \\ 3c - 8d = 0 \\ c + d + e = 1 \end{cases}$$

whose solution is $c = 2/3$, $d = 1/4$, $e = 1/12$.

Exercise 2.4. The community of warlocks, wizards, and magic practitioners is anxious to know the winner of the Elementalist Duel, which will be either of the powerful Absolom (A) or the great Balthazar (B). Assume that the payoff of either magician is $+1$ for winning and -1 for losing the tournament, respectively. The rules of the duel are simple: each elementalist will choose one elemental spell to face the opponent and cast it from his wand. This game is played very quickly without any chance of changing the spell. The spell can use one of four elements: Fire (F), Ice (I), Rock (R), and Wind (W). They are in the following relationships: Fire beats Ice, which in turn beats Rock; Rock beats Wind, which in turn beats Fire. If either magician chooses an element that beats the other, he will be the winner. Additionally, the slightly more powerful Absolom will also win in the case that they choose the same element. However, if they choose different elements that do not prevail over each other (such as Fire and Rock, or, Ice and Wind), the smarter Balthazar will ultimately be the winner. Finally, Absolom has been involved in a gruesome accident when an orphanage burned down because of a miscast "Fire" spell, so he has been forbidden to use this element by the Magician Guild, and Balthazar is aware of that.

1. What kind of game is this? Represent it in normal form. If you answered the question, you know you can only write Absolom's payoffs.

2. Write down a minimax optimization problem with the purpose of finding the Nash equilibrium of this problem. Write it from Absolom's perspective, with a_1, a_2, a_3 as the weights of its possible actions in his mixed strategies.

3. Solve the resulting minimax problem and find the strategy played by Absolom at the Nash equilibrium. [Suggestion: you can exploit that, for this problem, the minimax in mixed strategy must be $w = 0$.]

   [Bonus question: answer only after being sure about the rest]
   What is Balthazar's strategy at the Nash equilibrium?

Solution

1. The game is zero-sum, so Balthazar's payoffs are always opposite to Absolom. Absolom's pure strategies are $\mathcal{S}_{\mathtt{A}} = \{I, R, W\}$, while Balthazar's are $\mathcal{S}_{\mathtt{B}} = \{F, I, R, W\}$. You can write down only Absolom's payoffs in the following matrix.

		B			
		F	I	R	W
A	I	-1	+1	+1	-1
	R	-1	-1	+1	+1
	W	+1	-1	-1	+1

2. Absolom's mixed strategies are $a_1 I + a_2 R + a_3 W$. The minimax optimization problem can be written as:

$$
\begin{aligned}
&\max w \\
&\text{s.t.} \sum_{j=1}^{3} a_j = 1 \\
&\quad a_j \geq 0 \quad \forall j = 1, 2, 3 \\
&\quad \sum_{j=1}^{3} a_j u_A(\mathcal{S}_A^{(j)}, \mathcal{S}_B^{(k)}) \geq w \quad \forall k = 0, 1, 2, 3
\end{aligned}
$$

where w is a slack variable to be maximized, that can be at most equal to the maximin = minimax.

3. Observing that the highest possible value for w is 0 the last four equations can be rewritten as:

$$
\begin{cases}
-a_1 - a_2 + a_3 \geq 0 \\
+a_1 - a_2 - a_3 \geq 0 \\
+a_1 + a_2 - a_3 \geq 0 \\
-a_1 + a_2 + a_3 \geq 0
\end{cases}
$$

for positive a_j summing to 1. The first equation corresponds to $a_3 \geq a_1 + a_2$ while the third corresponds to $a_3 \leq a_1 + a_2$. Hence, $a_3 = a_1 + a_2$. Similarly, from the second and fourth equation we get $a_3 = a_1 - a_2$. Thus, the solution is easily found as $a_1 = 1/2$, $a_2 = 0$, $a_3 = 1/2$. This is the mixed strategy m played by Absolom at the Nash equilibrium.

[Bonus question] There are several mixed strategies $b_0 F + b_1 I + b_2 R + b_3 W$ that Balthazar can play at the Nash equilibrium. One of them is clearly $b_0 = b_1 = b_2 = b_3 = 1/4$. In general, the problem from Balthazar's perspective can be solved to find that all 4-tuples (b_0, b_1, b_2, b_3) are a Nash equilibrium as long as $b_0 \geq b_2$, $b_1 \geq b_3$ and $b_0 - b_1 = b_2 - b_3$, so also $(1/2, 1/4, 1/4, 0)$ or $(1/4, 1/2, 0, 1/4)$ would work.

Exercise 2.5. In a strategy tabletop game, the red player (R) is attacking the blue player (B). The rules of the game, known to both players, are as follows. Both players can choose out of three options, i.e., an integer number between 1 and 3 included, unbeknownst to each other. Then, the numbers are revealed and the winner is determined. If they both chose the same number, the winner is B and R loses. If the numbers chosen are off by 1 (regardless who chose the bigger), the winner is R and conversely B loses. Finally, if the numbers are off by 2 (i.e., 1 and 3, in any order) there is no winner nor loser. Assume the winner receives 1 zillion euros from the loser.

1. Formalize this as a static game and represent it in normal form.
2. Find out all the Nash equilibria of this game in pure strategies.
3. Find *one* of the possible mixed-strategy Nash equilibria.

Solution

1. The set of players is {R, B}. Their strategy sets are $\mathcal{S}_R = \mathcal{S}_B = \{1, 2, 3\}$. Payoffs are represented through the following matrix (the game is zero-sum, so only R's payoffs are shown):

R \ B	1	2	3
1	-1	1	0
2	1	-1	1
3	0	1	-1

2. The game has 9 possible outcomes in pure strategies, but none of them is a Nash equilibrium.

3. Because of Nash theorem, the game must have at least one mixed strategy Nash equilibria.

   You can look for a mixed-strategy NE where all the pure strategies 1, 2, 3 are played by both players with non-zero probability. If you call α and β the probabilities that R plays 1 and 2, respectively, then the probability of playing 3 is $1 - \alpha - \beta$. The conditions to be satisfied at the equilibrium are:

$$\begin{cases} u_B\big((\alpha, \beta, 1-\alpha-\beta), 1\big) = u_B\big((\alpha, \beta, 1-\alpha-\beta), 2\big) \\ u_B\big((\alpha, \beta, 1-\alpha-\beta), 1\big) = u_B\big((\alpha, \beta, 1-\alpha-\beta), 3\big) \end{cases}$$

   thus, plugging in the numbers we obtain the linear system

$$\begin{cases} \alpha - \beta = -\alpha + \beta - (1-\alpha-\beta) \\ \alpha - \beta = -\beta + (1-\alpha-\beta) \end{cases}$$

   whose solution is $\alpha = 2/7$ and $\beta = 3/7$. With analogous computations, the same probabilities are obtained for B. Note: this is not because of symmetry, as the game is zero-sum and therefore not symmetric. It is just a numerical coincidence. Hence, this Nash equilibrium in mixed strategies consists in both A and B playing 1 with probability $2/7$, 2 with probability $3/7$, and 3 with probability $2/7$.

This is the only NE, but proving it is actually not simple. Ideally, one should check all other mixture of probabilities. For example, consider a mixed-strategy NE where both players play only strategies 1 and 2 with non-zero probability, corresponding to considering the 2×2 minor of the matrix.

		B 1	B 2
R	1	-1	1
	2	1	-1

You can see that this reduced form of the game has an NE where, for symmetry reasons, both strategies are played with 0.5 probability. Unfortunately, this is *not* an NE according to the indifference theorem. An important condition (often forgotten by students) of the theorem requires that a player chooses its mixed strategy so as to make the other indifferent for all pure strategies belonging to the support of the other player's mixed strategy, *but also* strategies outside the support of that player's strategy must not get a better payoff. This is not happening if both players choose to play a mixture m of 1 and 2 with 50-50 probabilities, since they get an expected payoff of 0. Now, this value is the same of what they get by playing pure strategies 1 and 2 against m, but unfortunately strategy 3 gets a payoff of 0.5, thus representing a better unilateral deviation, which proves that m is not the best response to itself.

So one should go through all possible combination, which is tedious. Of course when analyzing a game like that, one can use some software tools available online, which would confirm that the previously found NE is the only one. Or, in this case one can use the systematic approach of minimax to find the NE.

An optimization problem can be written for the probabilities r_1, r_2, r_3 that player R chooses strategies 1, 2, or 3, respectively, as

$$
\begin{aligned}
&\max w \\
&\text{s.t.} \sum_{j=1}^{3} r_j = 1 \\
&\quad r_j \geq 0 \quad \forall j = 1, 2, 3 \\
&\quad \sum_{j=1}^{3} r_j u_{\mathrm{R}}(\mathcal{S}_{\mathrm{R}}^{(j)}, \mathcal{S}_{\mathrm{B}}^{(k)}) \geq w \quad \forall k = 1, 2, 3
\end{aligned}
$$

where w is a slack variable to be maximized, that can be at most equal to the maximin = minimax. The highest possible value for w is $1/7$, since the previously found NE achieves this value and the minimax theorem guarantees that this is the value of the game. Hence, the problem can be rewritten as

$$\begin{cases} -r_1 + r_2 \geq 1/7 \\ +r_1 - r_2 + r_3 \geq 1/7 \\ +r_2 - r_3 \geq 1/7 \end{cases}$$

for positive a_j summing to 1. As can be easily checked, all constraints are active (i.e., satisfied with equality), therefore confirming that the support of the mixed strategy contains all three pure strategies. Thus, the only solution for a mixed NE is $r_1 = 2/7$, $r_2 = 3/7$, $r_3 = 2/7$. In a sense, the situation is now symmetric for player B, not in the sense that the payoffs are the same, but that taking the dual problem of minimizing the value, which is B's objective, leads to the same values for the probabilities $b_1 = 2/7$, $b_2 = 3/7$, $b_3 = 2/7$ that are weighing B's strategies in its mixed strategy.

3

Sequential games

Sequential games with perfect information are a particular class of *dynamic games*.

3.1 Characteristics

In sequential games of n stages, players alternate in making their moves, that is, player 1 moves alone at stage 1, then at subsequent stages $j = 2, 3, \dots, n$, it is player $j = 2, 3, \dots, n$'s turn to move. In the case of perfect information, the player moves being informed about the choices made within all the previous $j - 1$ stages. This last element is key to understand the game: all moves are made with full awareness of the previous choices. Clearly, all players see the evolution of the game in its entirety, but this assumption of perfect information is actually only relevant to the players to move, since the players are also generally assumed to have perfect recall, so they do not forget past moves. Also, it is equally importantly assumed that all the players are rational and this information is common knowledge among the players.

Because of the properties mentioned above of the clearly specified order of moves and full information about the past history of the game, in a sequential game with perfect information the *information set* of each node is a singleton, which means that, when the gameplay dynamically evolves to a certain in-game situation, all the players (and most notably, the player to move) are perfectly informed about where the game is at. Thus, a sequential game is generally represented in extensive form by a plain tree without any possible confusion of the players to move; in other words, there are no dashed lines joining separate nodes of the tree. The extensive form may be preferred because it preserves the dynamic character of the game. The normal form can also be used, but its representation is generally not very compact, the reason being, since each information set is separate and only contains one node (the current one), one must specify one action per node in the tree, and the number of choices, which becomes combinatoric, explodes pretty soon.

The general scheme described above also includes games that apparently do not fit the definition. For example, a player may move multiple times at different stages i and $j \neq i$, that is, player i and j in the definition above can coincide. It is possible that just 2 players alternate in making choices, e.g., if all odd-numbered players and all even-numbered players actually coincide with players

1 and 2, respectively. In that case, it would simply be a game with two players deciding upon their choice after seeing each other's last move. Many tabletop games, played with an open and fully visible board, and without the involvement of hidden elements (cards) or random choices (dice), such as chess, checker, or tic-tac-toe fall within this category. In the extensive form, that would be represented as different layers of depth of the tree being similarly labeled, to describe that the player to move in a certain node has also made a choice in the past.

Also, many sequential games include an "early-termination" option where one player may have, among their choices, the one to conclude the game before stage n, so there are no more iterations and payoffs are immediately computed. Actually, this also fits the general definition above, if we allow for choice allowed to a player, corresponding to the early termination, to involve a subsequent gameplay where, no matter what the other players do, the final payoffs are always identical. This means that, in the extensive form, such a sequential game can be equivalently represented by an unbalanced tree, where one of the branches exiting a node of level $m < n$ reaches a leaf node where payoffs are computed. That would actually be equivalent to keeping the tree balanced but allowing for all the leaf nodes belonging to that subtree to bear identical payoff labels. Notably, the extensive form representation is not unique, as multiple extensive forms, as a matter of fact, correspond to the same actual game, as in this case.

3.2 Backward induction

Backward induction is a technique that can be used to solve games where moves unfold over time, in particular sequential games, for which it represents the instrument of choice for the solution. It corresponds to applying sequential rationality but, as the name suggests, it proceeds backwards, starting from the in-game situations where a last rational decision can be made. Once the uncertainty related to that decision is removed, the procedure goes on by considering the penultimate rational decision before that one, and so on.

This means that, if the game is represented as a tree (extensive form), backward induction starts by considering nodes at the end of the tree, such as those who are only followed by leaf nodes. Some branches coming out of these nodes can be removed by exploiting rationality, and subsequently the attention is moved to nodes up in the tree towards the root, which after the removal become closer to the end nodes.

This characteristic explains the name, since the strategic gameplay is planned by *inducing* further rational decisions in the tree, and doing so by going *backwards* in the order of the decisions to be made. This way, the best decision path is eventually found.

However, since backward induction eventually applies rationality to *every* single node in the tree where a decision can be made, the procedure does not only find Nash equilibria of the game, i.e., it determines optimal decisions along the equilibrium path, but also necessarily finds subgame perfect equilibria.

Indeed, the equilibrium path involves optimal decisions by the players, but at the same time the choices made off the equilibrium path are also fully rational, even though they may be cut off from being played by subsequent iteration of the backward induction.

In the language of mathematical optimization, backward induction corresponds to dynamic programming by applying the Bellman optimality principle, that is, the optimal path over a sequence of n steps is found as the best combination of an initial step and the optimal choice over the subsequent $n - 1$ steps. For this reason, backward induction over a game is essentially the same as the application of Bellman equations in a dynamic optimization problem.

The main difference of context is that, in an optimization context, dynamic programming is mostly a divide-et-impera paradigm to reduce a complex optimization into a set of simpler choices. In game theory, the local choices are already usually simple, such as a choice over a finite, and usually fairly narrow, set of alternatives. However, they involve *different* players and therefore backward induction heavily relies on the assumption of rationality of the players, as well as its being common knowledge among them. Since players are all rational, and also aware that all of them are, they can share the same conclusions about subsequent decisions to be made rationally.

As a useful criterion to approach practical cases, one must remember that the reduction in depth over the decision tree caused by taking the only best solution available does not always lead to a unique result, in general. However, if the final payoffs of all players in the leaf nodes are different, this is sufficient to guarantee the uniqueness of the subgame perfect equilibrium found by backward induction. Otherwise, a tie-breaker rule is generally required, to precisely determine what choice will be made by a rational player facing equivalent choices.

We remark that generalizations of the methodology behind backward induction, i.e., sequential rationality, allow to solve other special cases of games, where standard backward induction would seem impossible to apply. The first case is that of infinite-horizon sequential games with perfect information, where certain choices of the players lead to an early termination where the payoffs are computed but it is also possible that the subsequent stages go on indefinitely if the players make certain choices. Sometimes, one can cut this endless unfolding of the game if the payoffs keep decreasing (or conversely, predict that the game will not terminate if they grow indefinitely). In fact, perfectly rational players are immediately able to realize when they should stop sending the game to the next iteration.

An approach similar to backward induction can be applied to some cases where a finite-horizon dynamic game has sequential elements, but also imperfect information, meaning certain moves are simultaneous. In this case, sequential rationality can still help, for example in the case where simultaneous moves have a unique Nash equilibrium. In this case, rational players can anticipate that this will be the outcome, and the approach of backward induction can be applied to decrease the depth of the decision tree.

3.3 Subgame perfect equilibrium

The concept of subgame perfect equilibrium is a refinement of that of Nash equilibrium, meaning that all SPEs are NEs, but not necessarily the other way around. This is because a subgame perfect equilibrium is defined as a strategy profile obtaining a Nash equilibrium of every subgame of the original game. Since the whole game is a subgame of itself, an SPE must be an NE, first; but at the same time, it must satisfy more stringent conditions.

The existence of SPEs is guaranteed, as long as the players have perfect recall. If backward induction leads to a unique equilibrium path of the game, this is the only SPE of the game. Most commonly, one can compare the Nash equilibria found from a normal form representation, while usually only one of those is an SPE, if the game falls in the cases described above.

This implies that games usually have Nash equilibria that are not SPEs. The reason for this is related to some non-credible behavior of the players. Indeed, whenever an equilibrium is not subgame-perfect, there must be an illogical response of some of the players that violates the principle of choosing a best response. However, this is not noticed (and therefore the NE is legit) because this happens off the equilibrium path. Usually, this translates in a non-credible threat.

As an important notice, remember that the SPE is not defined by its equilibrium path only. Very often an erroneous solution to a request for an SPE is given by only specifying the final outcome of the game. While this is usually the most important aspect (since it ultimately tells how the game actually evolves), it is conceptually wrong to provide this as the answer to a question about the SPE, for the very reason that the main concern when finding an SPE is to find what happens off the equilibrium path, and to show that it still corresponds to rational choices. And clearly, there may be multiple NEs in a game, not all of them being subgame perfect, but in the end leading to the same equilibrium path.

Exercises

Exercise 3.1. A university student (U) is writing a project for the Game theory course; the project can have 3 levels of quality: 0 (worst), 1, or 2 (best), and this is up to U's decision. After U prepares and submits the project, the professor (P) will review it and evaluate whether it is successful (S) or not (N). If the project is evaluated as successful, the payoffs of both players are evaluated as $m - e_j$, where m is the mark awarded and e_j is the effort spent by player j. Given that the project's quality is x, with $x \in \{0, 1, 2\}$, the mark is $m = 18 + 6x$; the effort spent by U is an exponential function of x, i.e., $e_U = 4^x$; the effort spent by P is constantly equal to $e_P = 15$. If the project is not successful, the payoff of both players is 0.

1. Write this game between U and P in extensive form.
2. Write the game in normal form, and find out all of its Nash equilibria.
3. Find all the subgame-perfect equilibria of this game.

Solution

1. The extensive form is the tree depicted in Figure 3.1. First move is by U, who has 3 options to choose from: 0, 1, and 2. For each of the resulting branches, P has two choices, S or N.

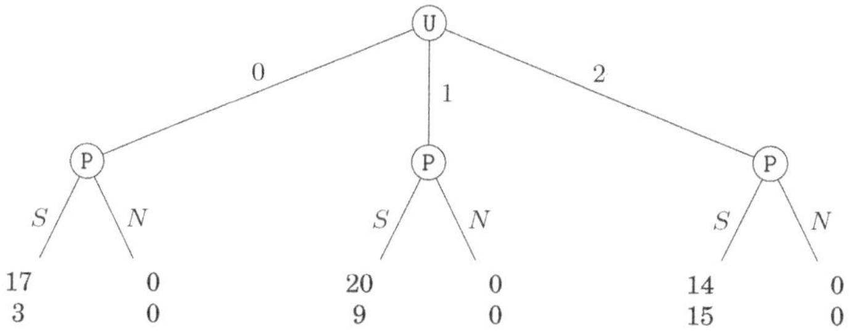

Figure 3.1: Extensive form of the game.

2. The normal form is a matrix (actually a bi-matrix) with 3 rows (because the first player U has 3 strategies) and 8 columns. The second player P has indeed 8 strategies because it must specify a binary alternative for 3 possible history of play, i.e., the choice of U. This results in:

C

B		SSS	SSN	SNS	SNN	NSS	NSN	NNS	NNN
	0	17, 3	17, 3	17, 3	17, 3	0, 0	0, 0	0, 0	0, 0
	1	20, 9	20, 9	0, 0	0, 0	20, 9	20, 9	0, 0	0, 0
	2	14, 15	0, 0	14, 15	0, 0	14, 15	0, 0	14, 15	0, 0

There are 7 pure strategy Nash equilibria: (0, SNS); (0, SNN); (1, SSS); (1, SSN); (1, NSS); (1, NSN); (2, NNS).

3. Of these Nash equilibria, the only SPE is (1, SSS), which can be seen via *backward induction*. The only rational choice for P is to say S to every project. The first step of backward induction is hence replacing P's nodes with the payoffs obtained from having P playing S.

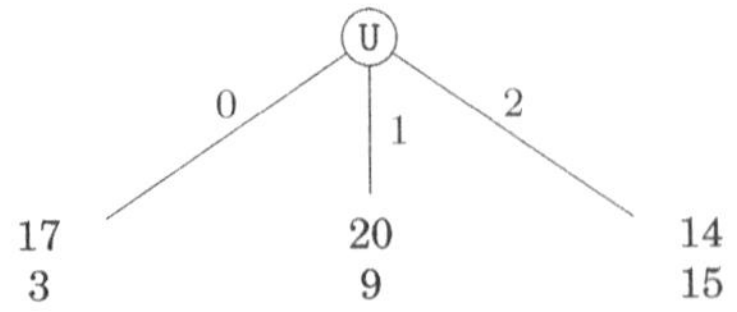

Knowing this, U will play the one that is best for him/herself, i.e., 1.

We can see that Nash equilibria (1, $\bullet S \bullet$) where P's strategy is not SSS are not sensible. In this case, P plays action N for some choices of U; the trick is, this can only happen when this is not the actual chosen project by U. But then, P is not playing rationally in the corresponding subgame (but luckily, this is off the equilibrium path). Then these are Nash equilibria, but are not subgame-perfect. Similar considerations hold for Nash equilibria (x, $\bullet N \bullet$) where $x \neq 1$. These are Nash equilibria where P would not accept project 1, which is not sensible, but luckily U chooses another project and P accepts it. Thus, a potentially illogical choice of P happens off the equilibrium path and therefore it is not captured by the Nash equilibrium (but it is instead visibile when considering the SPE).

This can also be understood as a non-credible threat. Imagine a strict professor that threatens to reject the project unless it is top quality; in other words, P threatens to play NNS – or better, the professor is *believed* by the student to be acting like this, because P's move is eventually unknown when U plays. Then, the student may be afraid that he/she has to deliver a high quality project, which seems to be the only way to be successful. Backward induction reveals this belief to be non-credible.

Side note: even though it may happen that professors have no incentive in rejecting a student's project, no matter how bad it is (after all, if they compel the student to do a new project from scratch, they have to grade one more project, which is against their utility), if you do a project in your game theory class you are encouraged to do better than the bare minimum!

Exercise 3.2. Miss One (1) and Mister Two (2) play the following game. They start with respective payoff $u_1 = +1$, $u_2 = -1$. At first, 1 can choose between two available moves, called A and B. Move A multiplies both payoffs by 3. Move B multiplies them both by 2 and additionally gives 5 more points to Mister 2 (so $u_2 = 2u_2 + 5$). 2 observes this choice of 1 and can either continue playing (C) or drop (D). If D is chosen, the game ends and the values of u_1 and u_2 are now the respective payoffs of the players. If 2 decides to continue, miss 1 will have another shot, and she can play again either A or B, with the same consequences of before. After that, the game ends, and the values of u_1 and u_2 are now the final payoffs of Miss 1 and Mister 2, respectively.

1. Write the game in extensive form.

2. Find out the backward induction outcome of the game.

3. How many strategies do 1 and 2 have, respectively? Write down their choice of strategies in the only subgame perfect equilibrium of this game. Are there other Nash equilibria that are not subgame perfect?

Solution

1. The extensive form is the tree depicted in Figure 3.2.

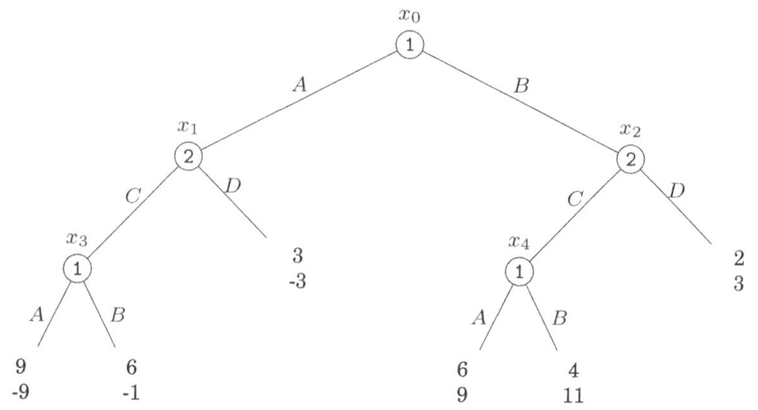

Figure 3.2: Extensive form of the game.

2. To apply backward induction, note that the last level of nodes is that below x_3 and x_4. In the last round, it is always more convenient for 1 to play A, thus x_3 and x_4 can be treated as final nodes with payoff $9, -9$ and $6, 9$, respectively.

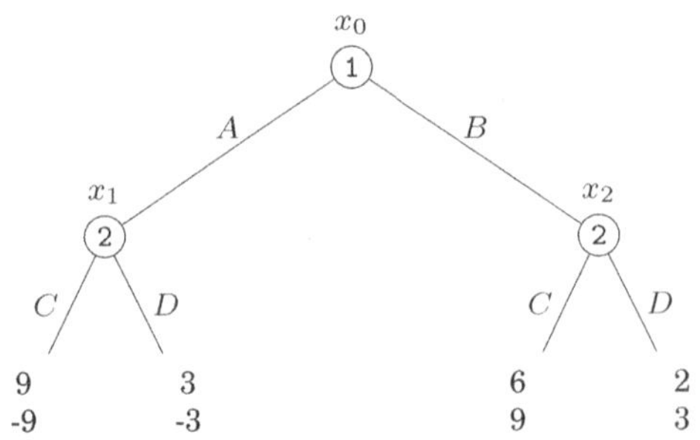

Now the last choice is always Mister 2's. At node x_1, he will choose D, which gives him a higher payoff, while at node x_2 he will choose C.

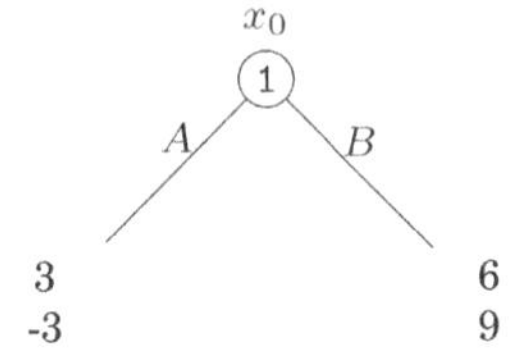

So now we are left with Miss 1's comparing two alternatives at node x_0, and the choice is between A giving an ultimate payoff of $3, -3$, or B giving an ultimate payoff of $6, 9$. Miss 1 chooses B and the final outcome gives payoffs $6, 9$.

3. Strategy-wise, 1 has two options to choose from at each of nodes x_0, x_3, x_4. Thus, she has 8 possible strategies. Similarly, 2 has two options to choose from at each of nodes x_1 and x_2. Thus, overall he has 4 possible strategies. We can write a pair of strategies as $((x_0, x_3, x_4), (x_1, x_2))$. The SPE outcome is what determined by backward induction, which dictates $x_0 = B$, $x_2 = C$, $x_4 = A$. Nodes x_1 and x_3 are off the equilibrium path, but in the SPE case they must be $x_1 = D$ and $x_3 = A$, thus the SPE is $((B, A, A), (D, C))$. There are clearly other non-SPE Nash equilibria. For example, we can change x_1 and/or x_3: for example, $((B, B, A), (C, C))$ obtains the same payoff of the SPE and thus no player has incentive to deviate from it; yet, it contains irrational choices at nodes that are never reached. In addition, there are Nash equilibria that are not SPEs because they have non-credible threats. One example is $((A, A, A), (D, D))$, where 1 always plays A and 2 always drops. Because of 2 always playing D, 1 has no incentive in playing anything else than A and vice versa.

Exercise 3.3. It is the discount sale season, and Lou (L) wants to go shopping. He thinks it is best to wait until the last three days of the discount sales, because prices are cheapest (denote these last three days as "Day 1", "Day 2", "Day 3"). On day 1, Lou asks Karen (K) to go with him. If Karen says yes (Y), they go shopping and the game ends. If Karen says no (N), Lou can either give up (G) or request (R) again the following day. On day 2, the same happens. On day 3, if K still says N, the game ends as well and does not go into a further day. If in the end K and L do not go shopping, both of their payoffs are 0. If they do, their payoffs are computed as $u_K = d$ and $u_L = 5 - 2d$, respectively, where d is the day in which they go. Note that these payoffs already include a discount phenomenon (only, L is eager to go as soon as possible, so the higher d the lower his payoff; while K prefers to go as late as possible, so her payoffs is kind of *negatively discounted*) so no further discount is to apply. All of this information about the game is common knowledge among the players.

1. Represent the game in its extensive form.
2. Find out the result of the subgame-perfect equilibrium of this game.
3. Find out another Nash equilibrium that is not subgame-perfect.

Solution

1. This is like a centipede game with extensive form shown in Figure 3.3.

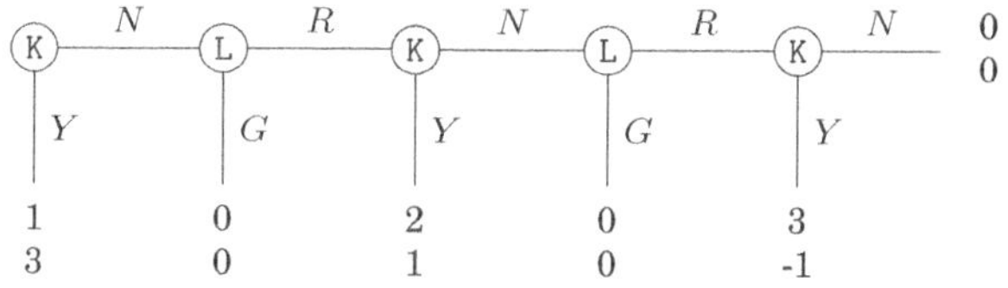

Figure 3.3: Extensive form of the game.

2. The subgame-perfect outcome can be found through backward induction. In her last choice, K prefers Y over N but this gives a negative payoff to L.

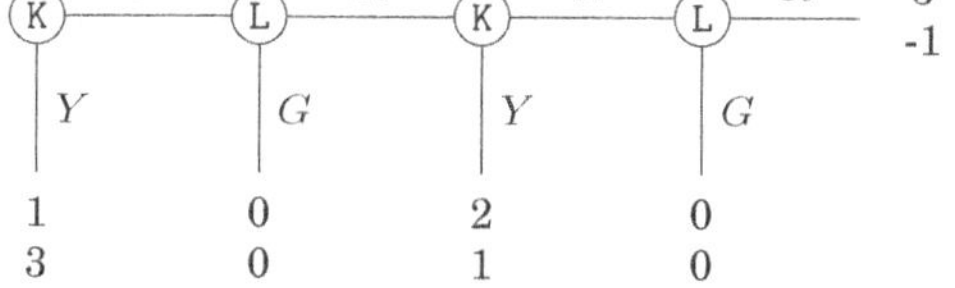

Anticipating this, L prefers to exit the game at his second choice and get 0.

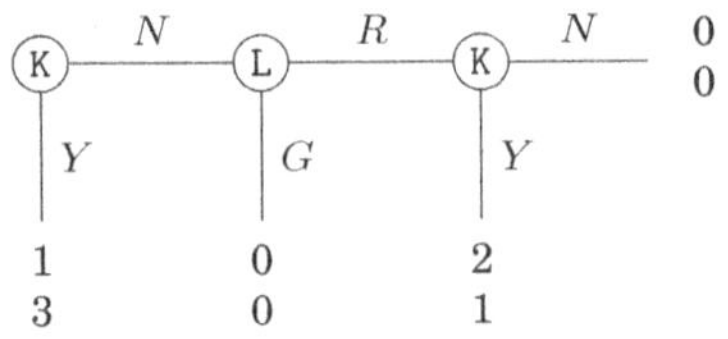

Then K prefers saying Y at day 2 and she can force this outcome (L prefers this to getting 0 and K prefers this over getting 1 by saying Y already at day 1). So, the result of the SPE is that players choose N, R, Y and the game ends with payoffs $(2, 1)$. To be more precise, one should also mention what happens off the equilibrium path. In the subgame where L can play at day 2, he plays G. And in the subgame where K can play at day 3, she plays Y. Neither of these happens, as on the equilibrium path the game is ended before that.

3. There are others Nash equilibria that are not subgame-perfect. For example, if L threatens to always play G, K will say Y immediately at day 1. So K and L playing (Y,Y,Y) and (G,G) is a Nash equilibrium for the game - as well as any other pair of strategy starting with Y and G, such as (Y,N,N) for K and (G,R) for L. They are all equivalent, since the game ends immediately, but they contain some non-credible threat by L who threatens to give up.

Exercise 3.4. Javier (J) and Krista (K) want to buy a house together with their joint savings and move with each other. They agreed on the following plan: during September, J will go browsing available houses: if he finds something that he likes, he will buy it for both. If J finds nothing good, K will go during October in search of a house, and similarly she will buy for both anything of her preference that is available; or, she can decide not to buy anything and leave the decision for J to make in November, which is the last month. If K has not found anything satisfactory during her turn, J will go searching for a house in November and buy the best option he can find, without the option to further delay. Basically, regardless of whom signs the deal with the real estate, the procedure ends whenever either J or K decides to buy. Also, note that houses stay available just for one month: if not bought when available, they are sold to someone else.
This is the list of affordable houses, together with J and K's utility if bought:

Name	Available during	J's utility	K's utility
Atlas heights (A)	September	10	5
Bellevue park (B)	October	15	28
Cicada gardens (C)	October	5	12
Daffodil square (D)	November	18	15
Ember lane (E)	November	8	35
Floral route (F)	November	40	10

All of these values and the rules above are common knowledge for the players. Denote with X the choice not to buy a house and go to the next round.

1. Write this game between J and K in extensive form.
2. Write the game in normal form, and find out all of its Nash equilibria.
3. Find all the subgame perfect equilibria of this game.

Solution

1. The extensive form is the tree depicted in Figure 3.4.
2. The normal form is a bi-matrix with 6 rows and 3 columns. In fact, player J has 6 strategies (2 choices in the first move and 3 choices in the second move). Player K only moves once (if at all) and in that case she chooses among three options, so she has 3 strategies. It may even be that her moves do not matter at all, but we need to specify them.

K

J		B	C	X
	AD	10, 5	10, 5	10, 5
	AE	10, 5	10, 5	10, 5
J	AF	10, 5	10, 5	10, 5
	XD	15, 28	5, 12	18, 15
	XE	15, 28	5, 12	8, 35
	XF	15, 28	5, 12	40, 10

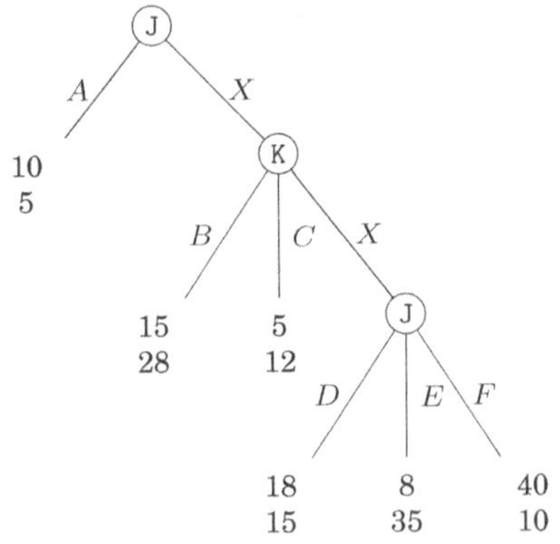

Figure 3.4: Extensive form of the game.

The game has the following 5 NEs: (AD, C); (AE, C), (AF, C), (XD, B), and (XF, B).

3. Indeed, there is only one subgame perfect equilibrium (XF, B), found as the outcome of backward induction. This SPE is unique as all the final payoffs in the leaf nodes are different, and the information sets are all singletons. To find it, observe that in round 3 player J will choose F that is the action giving him the highest payoff.

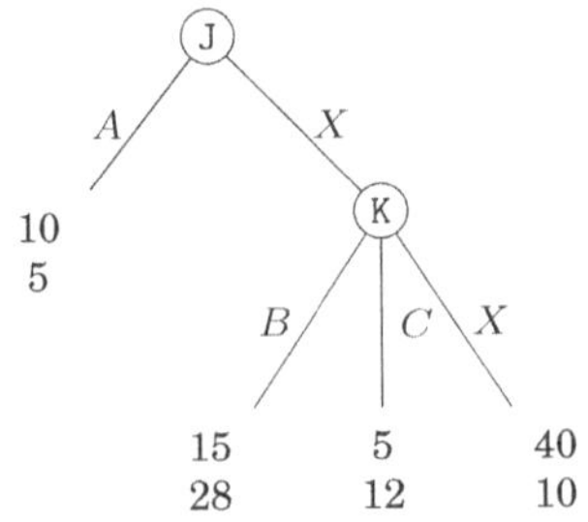

Knowing that, K prefers not to send the game to the third stage and terminates it earlier, since she can do that satisfactorily by choosing B and ending the game earlier. Anticipating that, J can play A instead but he is clearly worse off in doing so, thus J plays X in the first round instead.

An important warning: since the SPE implies that K ends the game during her turn and does not let J to perform his last move, one may be tempted to answer that the SPE is "J delays, and K chooses K." However, this answer would be *wrong*, since it does not fully declare J's strategy, which must be also specify what J does in the last round (even though it is not actually played). Indeed, notice that this description would also fit (XD, B) which is also an NE of the game, but not an SPE – since it implies a non-credible choice by J (even though it happens off the equilibrium path).

Exercise 3.5. Yvonne (Y) and Zoe (Z) are foreign sisters spending their summer holidays in Italy. They need to choose where to spend the famous "Ferragosto" night, and they have two options: the beach (B) or the discotheque (D). They decide independently and without telling each other, confident they will meet. However, Y has a car and drives to her location of choice; also, if Y does not find her sister at her chosen location, she (and only she) has one further option: to change location (C) to reunite with her sister, or stay (S) where she is. Instead, Z takes a one-way bus trip and is forced to remain at her location. The payoffs are evaluated as follows. Y prefers B, who brings her utility 4, while D gives her 1. Z's preferences are reversed and D gives her 7, while B gives her 1. Moreover, if they celebrate Ferragosto together, they both get their utilities increased by 5. Finally, Y's final utility is decreased by 3 if she needs to drive her car twice, i.e., to play C in the second round to reach her sister (staying where she is already does not change the utility from the first round). For example, if Y plays B and Z plays D, Y gets a second move; if this move is C, the final respective utilities are $u_Y = 3$ and $u_Z = 12$. All of the above information is common knowledge among the sisters.

1. Write down the extensive form of this game.
2. Find the outcome of the subgame-perfect equilibrium of this game.
3. The extensive form includes three points where Y moves, thus her strategy must be a triple of binary variables. What is the strategy played by Y in the subgame-perfect equilibrium?

Solution

1. The extensive form is the tree depicted in Figure 3.5. Y moves first choosing B or D, but her choice is actually unknown to Z so the two nodes are joined in the same information set.

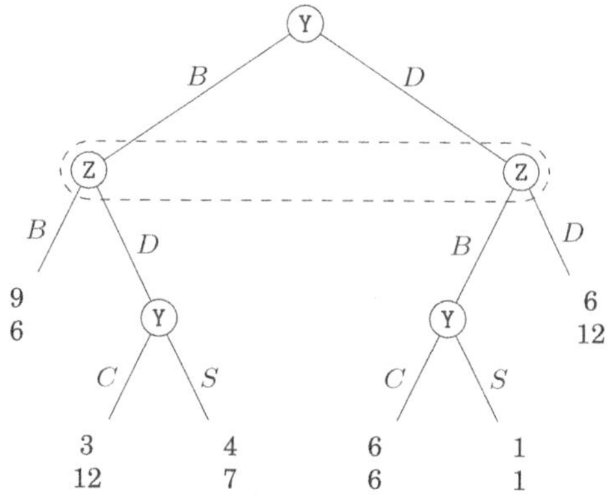

Figure 3.5: Extensive form of the game.

2. The outcome of the SPE can be found through backward induction. If Y plays a second round, she will choose S in the left-hand node (the one following (B, D)) and C in the right-hand one (following (D, B)). This result in the following reduced game, in normal form.

		Z	
		B	D
Y	B	9, 6	4, 7
	D	6, 6	6, 12

Clearly D is a dominant strategy for Z and the resulting NE is (D, D). The outcome of the SPE is thus both Y and Z playing D in the first round with final payoffs $(6, 12)$.

3. The strategy played by Y at the SPE above can be described as "DSC". The first binary value is a choice between B and D and describes what done at the top node. The second and third variables are what done at the nodes below and can be either C or S.

Exercise 3.6. Wife (W) and husband (H) share the chores of the house. Due to her working shift from 10:30 am to 18:30 am, W has two opportunities to do something, at 7:00 am and 7:00 pm. H instead works two shifts, early morning and afternoon till late evening, so he has only one opportunity to contribute, at 2:00 pm. There are two particular chores in the family: fix some food (F) or clean the house (C). If either family member performs a chore, he/she pays an individual cost, but *both* members receive an identical benefit: F gives benefit 30 but costs 10, C gives benefit 50 but costs 20. Moreover, any of these actions performed by W in its first opportunity costs 10% more but also gives 10% more benefit to both players (e.g., F costs 11 but gives benefit equal to 33). Alternatively, any player at his/her turn can "do nothing" (N) instead of choosing either chore. This can be done at any round, and gives individual benefit 10 only to the member choosing it (the social benefit is of course 0). Also, a chore cannot be repeated in the day (if food is already prepared in a previous round, the current action taker can only clean the house or do nothing). Action N can be repeated through all the rounds.

1. Write down the extensive form of the game.
2. How many subgame-perfect equilibria are in this game? Also, name the process on how do you find them.
3. Actually find them.

Solution

1. The extensive form is the tree depicted in Figure 3.6.

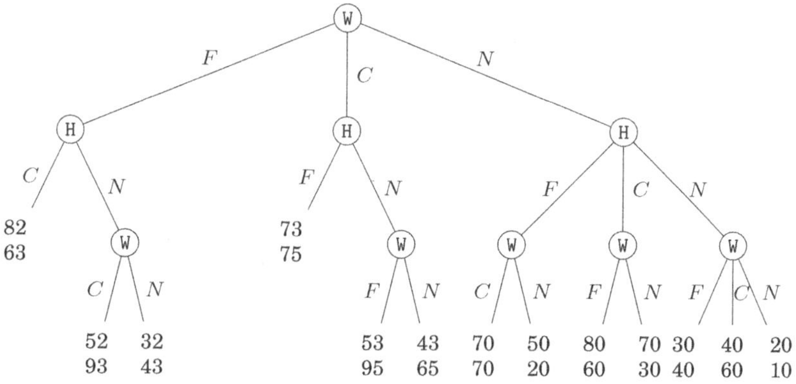

Figure 3.6: Extensive form of the game.

2. There is a single subgame-perfect equilibrium of this game, and the way to find it is called *backward induction*.

3. On the first round of backward induction, W is to play and all the nodes in the last level have at least one alternative – among the leaf nodes with the same parent – where the first player has a higher payoff. Now, the tree is left without these pruned leaf nodes, and therefore consists of three subtrees of F, C, and N.

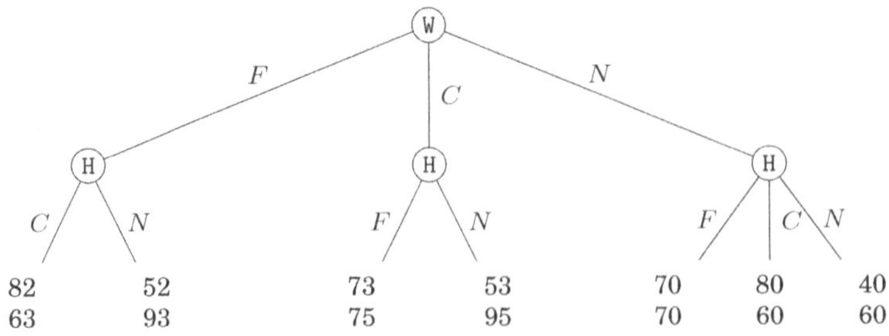

On the second round, H chooses among the new leaf nodes the ones that give him the highest payoff.

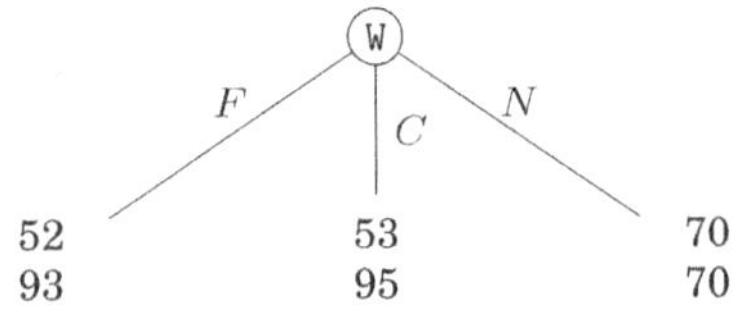

The last round consists in W choosing among F, C and N. She chooses N, since it is the option that maximizes her utility. Therefore, the outcome of the subgame-perfect equilibrium is W doing nothing (N) in the first round, H fixing food (F) in the second round and W cleaning (C) in the third round. This is actually only the equilibrium path: the subgame-perfect equilibrium must specify the strategy of the players in all the 9 subgames, even in the ones that are never reached. The joint strategy that describes the behaviour of the players in the various subgames can be written as

$$(s_W, s_H) = (NCFCFC, NNF),$$

where the strategy s_W is specified starting from the root and then for all the other nodes on the last level of the tree, from the left to the right, while s_H is specified for all the nodes in the central level, again from the left to the right.

Exercise 3.7. Two children named Anita (A) and Ben (B) are siblings. One afternoon, they are playing together when they both decide they want to get some candies without their parents to know. There are two cupboards with candies in their house, one in the closet (C), with 5 candies, and another in the dining room (D), with 10 candies. They will simultaneously choose one room each where to rush and grab candies. If they choose different rooms, each of them can get all the candies from the respective cupboards. If they go to the same room, they will have bargain on how to share the candies they find there. The way they will share the candies is that Anita will propose a split (denoted by the integer number of candies she will get; the proposal must include a positive number of candies for both siblings), and Ben can only agree with the proposal, or cry to complain. If Anita's proposal is accepted by Ben, the siblings will split the candies as that. Otherwise, Ben's complaints will cause the parents to notice them, and they will end up without any candy. Assume that the payoffs of the two children are equal to the number of candies they get, and that all of the aforementioned rules (including: the splitting rules, how many candies are in a room, that Anita is allowed to propose a split and Ben can only accept or refuse) are accepted by and known in advance to both of them.

1. Represent this game in extensive form.
2. Apply backward induction to reduce this game to a static game of complete information (i.e. single-shot): give its normal-form representation and find all its Nash equilibria.
3. Find the subgame-perfect equilibria of the original game and (if it exists) a Nash equilibrium that is not subgame-perfect.

Solution

1. The extensive form is the tree depicted in Figure 3.7. First (root node) is Anita's (A) move between C or D. Then Ben (B) will also choose between closet (C) or dining room (D) without knowing what A did. If the choices are (C,D) or (D,C) the game ends, with respective payoffs equal to (5,10) and (10,5) - A's payoff written first. If they both choose C or both choose D, they will play an Ultimatum-like game, with A being the proposer and B being the responder. Thus, A proposes a split and B can say Yes (Y) or No (N), in the former case the split also matches the ultimate payoffs they get, otherwise the payoffs are (0,0).
2. By backward induction, it is easy to see that all nodes where the ultimatum game is played, lead to a minimal offer of 1 candy from A to B, which is accepted (saying N is a non-credible threat). Thus, the normal form of the static game is:

		B	
		C	D
A	C	4, 1	5, 10
	D	10, 5	9, 1

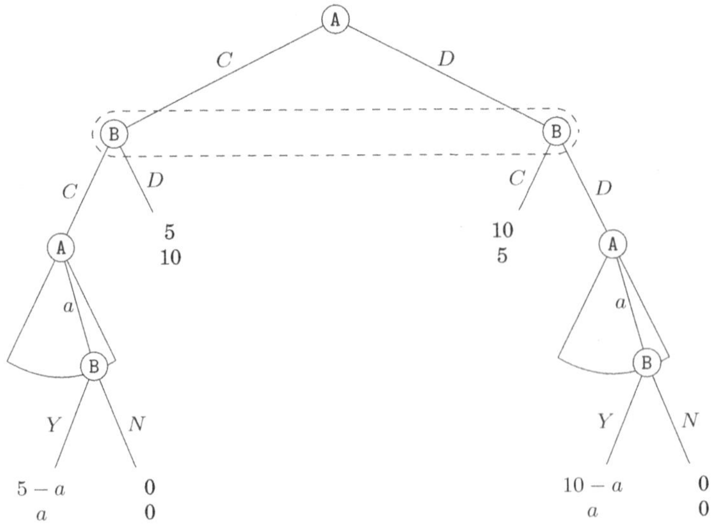

Figure 3.7: Extensive form of the game.

3. In this static game, strategies are just actions, and D is dominant for player A, thus C is dropped from her moves, leaving B's best option to play C. This means there is one Nash equilibrium in this game: (D,C), found by Iterated Elimination of Strictly Dominated Strategies.

   Therefore, there is only one subgame-perfect equilibrium in the original game, which includes: for player A, to play D and to offer split 4 or 9, respectively, if the first round ends with outcome (C,C) or (D,D), respectively; for player B, the SPE strategy is to play C and to accept any split proposed in the second round.

   The game has also other Nash equilibria that are not subgame-perfect. For example, a Nash equilibrium can be with A's strategy that is to play C and to keep some value $a > 1$ for herself in the case a second stage of the game (the bargaining part) is played. B's strategy can be to play D and, if a second round of game is played, to accept the split only if A gets 1 for herself. This is a Nash equilibrium (B does not want to deviate, A can deviate but in this case B will reject her proposal and she will get 0) but not an SPE: B's threat is non-credible as he will get payoff 0 too.

Exercise 3.8. Jeremy (J) and Kayla (K) are working on a project together. They agreed on the following plan: J starts working on the project on day 1, then at the end of the day he can either submit (S) the project or hand it over (H) to K, who will work on the project on day 2. At the end of day 2, K faces the same choice: either submit the project or hand it over to J for day 3, and so on, with J that can work on the project on odd days, and K on even days. When the project is submitted, whoever does the submission gets the full value of the project as a reward, and the other player only gets 60% of it. The value of the project is 2^t, where t is the current day (this means that every working day spent by J or K on the project, doubles its value). Also, the deadline for the project is day 10, after which the project value drops to 0. Everything said above is full knowledge for both players.

1. Describe this game in extensive form.
2. Consider the subgame-perfect equilibrium of this game; at which day is the project submitted?
3. Describe another Nash equilibrium of this game that is not subgame-perfect.

Solution

1. The extensive form is the tree depicted in Figure 3.8.

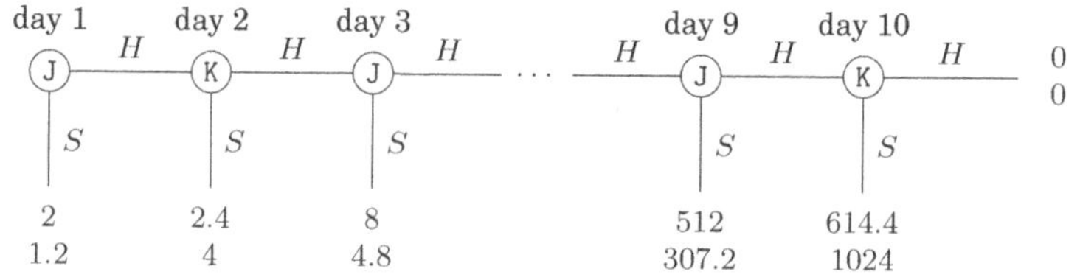

Figure 3.8: Extensive form of the game.

2. The last exit node is the one with highest payoffs for both, and the players can rationally anticipate that. So this is the subgame-perfect Nash equilibrium i.e., J always playing H and K playing "H only if the round is less than 10".

3. Another NE would be, for example, J playing H only if the round number is less than 9 and K playing H only if the round number is less than 30 (of course other numbers would work as well). In this case, any deviation by K cannot improve her payoff (she can only stop K from reaching round 9 but this gives her a lower payoff. Also J does not want to deviate since this means either to end the game earlier (lower payoff for both) or go beyond the deadline (payoff 0 for both). The trick here is at this non-subgame-perfect NE, K is keeping the irrational behavior of letting the deadline expire. This however never comes into play as J stops the game earlier.

Exercise 3.9. A university student association consists of a Sorority (S) section and a Fraternity (F) section; female members belong to the sorority, male members to the fraternity. The semester break is approaching: it is a period of 6 weeks (from week 0 to 5) when the association can throw parties, and they do so according to this rule: every week, the sections take turn in deciding whether to organize a party P or not N. As long as one section chooses P, they throw a party for all members of both sections and the week after is the other section's turn to decide. If one section chooses N, both sections agree that parties are over for that semester break. The first decision is assigned to section S at week 0. Throwing a party implies a cost 10 for the organizing section, however, it is also enjoyable for *both* sections, so they both get a positive *fun* value. The ultimate utility of the sections is their net benefit (fun minus cost). The fun for the party at week 0 is 25. Yet, throwing more and more parties clearly give a diminishing fun; thus, apply a discount factor of $\delta = 0.6$ for every party organized at week $j > 0$ (i.e., the fun value of a party at week j is $25\delta^j$). Note that, at week 6, S is forced to play N since the semester break is over.

1. Write the game in extensive form.
2. Find out the backward induction outcome of the game. Prove it represents a subgame perfect equilibrium and discuss if it is unique.
3. Discuss what would happen if there is no limitation to the length of the semester (and thus the length of the game), i.e., the students can go on partying to an arbitrarily large number of weeks. Can you find a subgame perfect equilibrium for this case?

Solution

1. The game is like a centipede game, whose extensive form is the tree depicted in Figure 3.9. The payoffs for player F are on the top, and on the bottom for player S.

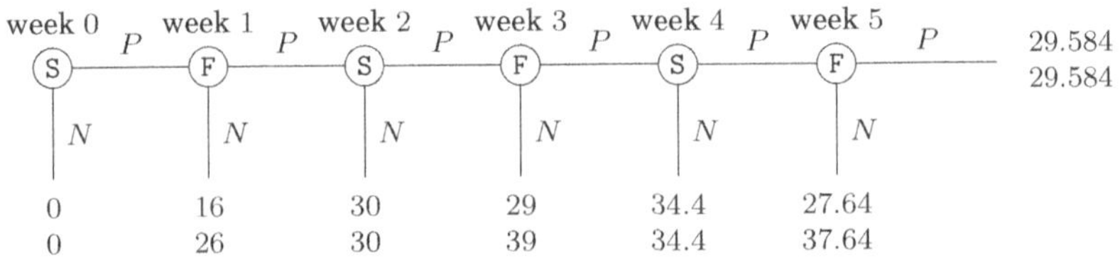

Figure 3.9: Extensive form of the game.

2. To apply backward induction, note that the last level of nodes is actually week 5, since both choices of F directly lead to a leaf node. Since $37.64 > 29.584$, F prefers N. But actually this means that at week 4, S must choose between P leading to a leaf node with payoffs $27.64, 37.64$ or N leading to a leaf node with payoffs $34.4, 34.4$; thus, S chooses N. Anticipating that, it is

immediate to discover that backward induction implies that both players choose (P, P, N) as their strategy. To be nitpicking, there is a subtle difference between the equilibrium path and the full SPE. The equilibrium path implies that S starts playing P at week 0, and so does F at week 1. They keep throwing parties until week 4, when S plays N thus ending the game. However, the full specification of the SPE must also include that, if given the chance, F would have played N too at week 5 – the thing is, this does not happen because the game ends earlier. However, only if F chooses N for week 5, they play a rational choice in every possible subgame. Finally, also notice that, since all terminal payoffs are different, theoretical conditions guarantee that the SPE is unique.

3. If there is no limit to the weeks (i.e., the branches in the tree from the side of P), backward induction cannot be applied as there is no end node. Still, we can apply sequential rationality inspired by the previous result. Specifically, the game has the same (only) subgame perfect equilibrium, because rational players are immediately aware that the game cannot go beyond week 4. Indeed, the first player S is to move during even-numbered weeks, and it is evident (to us, but also to the rational player S) that the achievable payoffs in nodes deeper than week 4 are always lower than what can be achieved by playing N at week 4. Similarly, the second player F is to move during odd-numbered weeks and their payoff cannot go beyond what achieved by playing N at week 5. Thus, the game is bound to be finished before week 6 even without the finite horizon imposed by the hard termination rule of the semester.

Exercise 3.10. Amelie (A) and Bernard (B) are doing a joint project together. They are both rational players and they know it. At day 1, the project has an initial value of v_1. Starting from day 1, they can either unilaterally call that the project is ended (C) or keep working on the project together (K). However, only one player can make such a decision. Amelie has the right to choose on odd days, while Bernard can decide on even days; this means that Amelie decides first at day 1, Bernard at day 2, and so on. If the decision at day j is to keep working on the project, the choice is deferred to the next day, where the other player decides; meanwhile, the value of the project is discounted by a factor $\delta \in [0, 1]$, so that at day $j+1$ the value is $v_{j+1} = \delta \cdot v_j$. As long as they keep deferring the decision, the game goes on forever. When the project ends, they both collect some payoff. Whoever called for the project to end at day j gets $0.4v_j$. The other player gets $0.6v_j$.

1. Draw the extensive form of this game, (write 4 days, then put some dots).
2. Prove that there is a unique SPE where the project ends at latest at day 2.
3. Find the values of δ for which the project will ultimately be ended by Amelie (the complementary values are those for which it is by Bernard instead).

Solution

1. The game is like a centipede game, with C on left (down) branches and K on right (horizontal) branches. The game stays on forever if a K-branch is chosen. Any choice of C ends the game, and final payoffs are $0.4v_j = 0.4\delta^{j-1}v_1$ and $0.6v_j = 0.6\delta^{j-1}v_1$ if j is odd; payoffs are swapped if j is even.

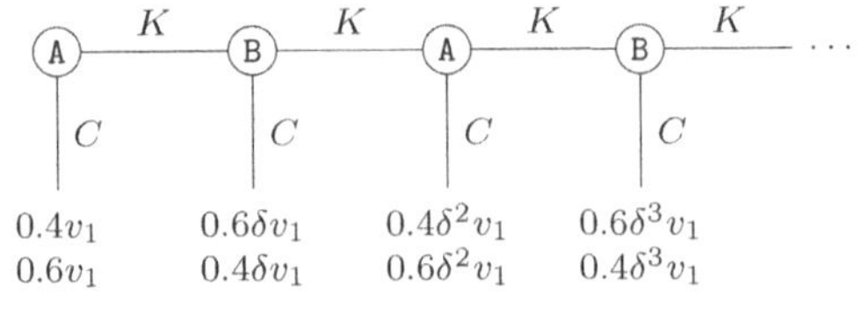

Figure 3.10: Extensive form of the game.

2. This game is bound to end at round 2 because both players know that at day 3 the game will be back to day 1, only with discounted payoffs. Rational players are immediately able to anticipate this outcome and they will not allow the game to reach round 3. Thus, it is either Amelie ending the project at round 1 or waiting for Bernard to end it at round 2. Notice that this resembles the theoretical principle of the dynamic game of *bargaining*, which can be proven to have a unique SPE even without resorting to backward induction, but just relying on the rationality of the players.

3. Since Amelie can anticipate Bernard's move, she compares $0.4v_1$ (what she gets at round 1) and $0.6v_2 = 0.6\delta v_1$, what she gets at round 2. Amelie chooses C if $\delta > 2/3$.

Exercise 3.11. A gang of pirates has n ranks from 1 (the ship's boy) to n (the captain). After a raid, they share a treasure. Pirate with rank $k+1$ keeps an eye on pirate k to see whether he gets a bigger share than he should. The game starts when pirate $k = 1$ (the ship's boy) realizes that the treasure contains an extremely valuable pearl that has fallen far from the stash: he considers whether to take the pearl for himself (P) hiding it in its pocket or do nothing (N). Doing nothing ends the game with the pearl being unnoticed and unassigned. However, if pirate k takes the pearl, pirate $k+1$ will notice it; now, pirate $k+1$ may consider to kill him and keep the pearl for himself (P), or do nothing (N). If pirate $k+1$ does nothing, pirate k is left alive with the pearl – a very good outcome. If pirate $k+1$ kills pirate k and takes the pearl instead, this is spotted by pirate $k+2$ that now faces the same choice: whether to kill pirate $k+1$ and keep the pearl for himself (P), or to do nothing (N). This means that k is replaced with $k+1$ and the game continues up to the captain. For every pirate, the top preference is to stay alive and have the pearl; after that, they all prefer being alive without the pearl than to be killed.

1. Consider $n = 5$. Choose appropriate utility values for the outcomes and draw the extensive form of the game.
2. Consider $n = 5$. Solve the game by finding its subgame-perfect outcome.
3. Consider $n = 8$. Does the subgame-perfect outcome change, and why?

Solution

Say utility of the pearl is 1, utility of being killed is -1, and utility of getting nothing but being alive is 0. This does not imply that pirates must be daring hotheads with low evaluation of their own lives, even though this is probably part of their lifestyle. Actually, it is only important that these three values are chosen in increasing order as specified by the text, so any other choice with the same preference order would also work.

The game is a tree where the root is player 1's move (this is level $\ell = 1$ of the tree), split into branches labeled P and N. The latter ends the game, with all payoffs from position ℓ up to n equal to 0, position $\ell - 1$ (if available) with payoff 1 and positions $< \ell - 1$ (if available) with payoff -1. Branch labeled with P sends the game to the level below. If P is played at the last node of the tree, payoffs are $(-1, -1, \dots, -1, +1)$.

For every value of n, player n who is the last to move, always chooses P resulting in $(-1, -1, \dots, -1, -1, +1)$ over N that would give $(-1, -1, \dots, -1, +1, 0)$. In other words, whenever the captain has the option to take the pearl for himself, he will gladly do so, as he faces no retaliation by anyone else. Thus, he will be the only one left alive, with the pearl. Remember, pirates are the quintessential game theory players, as they are definitely selfish and do not care about the utilities of others.

However, we can apply the power of backward induction to our reasoning, by considering player $n-1$ (the first mate) who is a rational player and can conclude immediately whatever discussed above. Thus, player $n-1$ always chooses N over

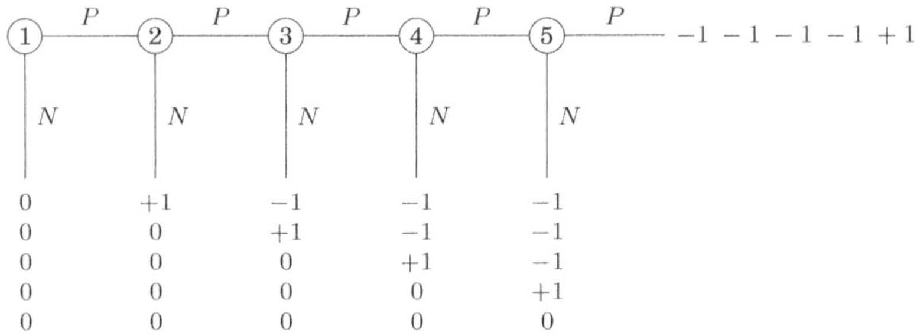

Figure 3.11: Extensive form for $n = 5$.

P, since this result in payoff $(-1, -1, \ldots, +1, 0, 0)$. This is not the best possible outcome for player $n-1$ (he will not get the pearl) but at least he will end the game alive, whereas sending the game to the last move by player n would surely mean his own death, and player $n-1$ does not like that. Notice that in this case the pearl will end up in the hands of a pirate with rank lower than the top.

Seen otherwise, if there is a single pirate ($n = 1$), he takes the pearl. If there are two pirates ($n = 2$), backward induction implies that the first pirate chooses not to take the pearl. If there are three pirates ($n = 3$), the second (also the second-last) does not take the pearl. Now, what should the first pirate do? He can apply all what said until now and realize that the second pirate will not take the pearl, so the first pirate can safely play P.

Iterating either version of these reasonings, it can be seen that, if the number n of pirates is odd, the outcome is that the first plays P (i.e., he takes the pearl), and the second plays N (i.e., leaves him alone), to avoid being killed, thus ending the game. If n is even, the first pirate plays N (i.e., does not take the pearl) for fear of being killed and the game ends.

Thus, there are actually two different outcomes of backward induction, depending on whether n is odd or even. In the text, the scenarios proposed include $n = 5$ and $n = 8$, but all that matters here is that 5 is an odd number, and 8 is even instead. For an odd number of pirates, the outcome is that the ship boy (pirate number 1) will take the pearl and pirate number 2 will pretend not to notice, to stay alive. If the pirates are in even number, pirate number 1 will not take the pearl (and so nobody takes it).

For what concerns the formal representation of the strategic choices of the players, they can be written as a single action, since all players have only one chance to move, and this is P or N for player j if $n-j$ is even or odd, respectively. Notice that this fully specifies the SPE even though many pirates will not have a chance to play. In particular, pirates with rank $n > 1$ come into play only after killing their subordinates, so they can still plan an action to play, and this is done according to the same principles as player 1's, since choosing P implies to be the pirate with lowest rank still alive.

4
Lotteries and Other Variations

In this chapter, we review some theoretical extensions that can occasionally take place in games. These do not represent very advanced theoretical concepts, and their presence in a game is certainly not standard. Whenever they are present, they should not be regarded as difficult hurdles. Most likely, they can be connected to other approaches previously discussed in other chapters. Thus, the way to tackle them is to "unplug" them from the game or to trace them back to another standard configuration.

4.1 Lotteries

Usually, rational players do not like random events, since they make the predictions about future evolutions of the game more difficult. However, in certain cases it is unavoidable to include some random externalities in a game. This is especially true in economic and social sciences, but can also be the case for information engineering scenarios, think for example of the transmission over a wireless channel whose attenuation is a random parameter.

For the sake of simplicity, discrete random variables are often considered, in line with the games usually having finitely many strategies and so on. The discrete probability distribution of such random variables is often referred to as a *lottery* and the actual value is said to be determined by a virtual player called Nature. We will see in the last chapter how this specific situation can be declined for Bayesian games, where a random choice of the types for the Bayesian players made by Nature is the spark that ignites the evolution of the game. Also, every deterministic outcome can also be seen as a *degenerate lottery*, where only one of the events happens, with probability 1.

While there are several possible approaches to compare uncertain outcomes, especially related to human behavioral studies, most technical applications of game theory find it convenient to take expectations and apply *expected utility hypothesis* to make such comparisons. This is true not only for game theory applications that do not involve human beings (but rather algorithms, intelligent systems, or transmission devices in a network, whose rationality is hardly questionable) but also in relationship to the theory of *vNM utility*, based on a theorem by von Neumann and Morgenstern. This theorem shows that the only way to satisfy four sensible axioms of comparisons among lotteries is to translate the same comparisons to the expectation over the lotteries of the value of the utility functions.

In other words, whenever a game dictates a choice by the virtual player Nature, rational players consider the resulting lottery as equivalent to a deterministic outcome equal to the expected value. When solving such a game, every lottery can be replaced with an equivalent degenerate lottery achieving the expected value, thus removing every random element. From the perspective of solving the game, this is neither really surprising nor puzzling, usually. Indeed, many other procedures imply to take expected values of the utilities (such as mixed strategies, or Bayesian games) and this is most likely based on the very same theoretical reasons of the aforementioned expected utility hypothesis. Still, the procedure is fully consistent as nested expectations are linear and just result in a single big expected value.

4.2 Tampering with elections

A side topic related to game theory is the declaration of preferences, which is usually represented through utility functions. But actually, this is just one possible way of expressing preferences in a quantitative form. Alternatively, one can explicitly list the preferences for all the alternatives from best to worst (or the other way around). As an important application especially to political systems, game theory is also able to investigate the issue of *strategic voting*, which essentially means that some electors lie about their preferences in order to play the system towards choosing a better candidate.

Even though this is a really common scenario during elections, where even the candidates themselves promote the electors of supposedly weaker candidate "not to disperse their votes" and choose them instead, this is by no means limited to that case, and in principle can be applied to any system where multi-objective optimization is performed (i.e., multiple and possibly contrasting criteria are used to decide what is best out of multiple choices), which is a relevant issue in artificial intelligence or cross-layer optimization.

It may seem paradoxical that changing a preference from truthful to insincere improves the result for a voter, since it violates monotonicity and/or transitivity of preferences. The fact is, the aggregate of multiple preferences is not transitive in general, due to the presence of the so-called Condorcet cycles, i.e., situations where a crowd of voters prefers **B** to **A**, **C** to **B**, but also **A** to **C**. In addition, it can be proven that no electoral system (or more in general, a way to aggregate preferences) is immune from other paradoxes that would incentivize strategic voting.

4.3 Stackelberg games

This is a fancy name for transmuting static games of complete information into dynamic games. In practice, this is not a separate topic, since it is fully covered by the case of sequential games discussed before. However, it is sometimes encountered in information technology applications, possibly because the name of German scientist Stackelberg sounds very science-y.

A Stackelberg game is nothing but a simple sequential game, but represented through a bi-matrix that resembles the normal form of a static game. In reality, this is not the correct way to put a dynamic game in normal form. As discussed before, a dynamic game generally has a much more complex normal form due to the combinatorial explosion of the strategies available to the players. However, a Stackelberg game is just meant to follow a very simple game dynamics, and therefore a slight abuse of notation is tolerated.

To give an example of the representation of a Stackelberg game, consider the following bi-matrix of payoffs.

L \ F	C	D
A	4, 3	-2, 0
B	0, 0	1, 1

The players are usually called Leader (L) and Follower (F), which reflects their order of play. Thus the previous table is not to be read as a normal form, in which case the game would have three NE (two in pure strategies plus another one in mixed strategies, since it would be a coordination game), but rather as corresponding to the following decision tree, which is the proper *extensive form* of the game.

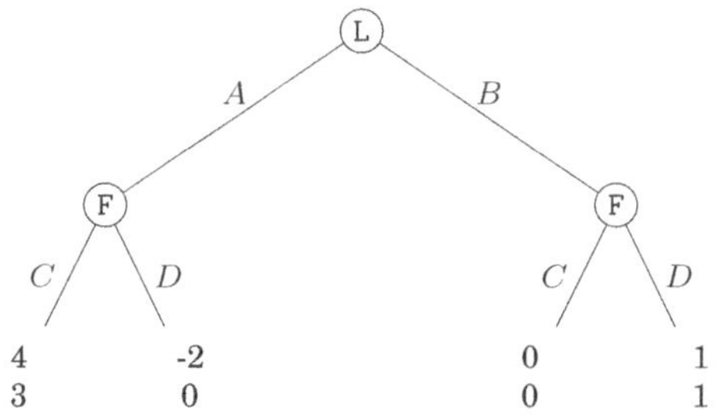

Most notably, the game has a unique "solution," often referred to as the Stackelberg equilibrium, which is actually nothing but the result of backward induction (and therefore an SPE). In practice, however, Stackelberg representations are not very rigorous and just give the equilibrium path. In other words, it is common practice to say that the Stackelberg equilibrium of the previous game is (A,C), even though the proper SPE notation, i.e., (A,CD), is more rigorous and must specify *two* actions in the follower's strategy (F must plan a move in two information sets, both singletons).

In short, Stackelberg games are not a separate theoretical topic, but they just correspond to comparing static and dynamic games and possibly drawing design conclusions, such as whether it is worth that a player of a 2-person game acts as a leader and communicates its choice to the other, who is forced to become a follower. This is often related to whether the "Stackelberg equilbrium" is better than the regular NE. In the example made before, an engineer would surely argue that from a practical standpoint, the Stackelberg approach avoids ambiguity on what is the final result, since nstead of three equilibria, we only have one, which is also giving the highest payoffs.

For this reason, Stackelberg equilibria are often popular in some research fields like network engineering, since they obtain a more definite result that can be better for algorithmic reasons. Yet, the reason why they achieve this result is not very deep, it is simply thanks to the good properties of backward induction, which is the real solution instrument here.

Exercises

Exercise 4.1. Students Luca (`L`) and Marco (`M`) are enrolled in the same degree, which involves a mandatory internship; they can spend it in either a company or a university lab. Students are allowed to apply to *only one* company. If the company approves the application, the student can spend the internship there. Otherwise, the internship must be done in a university lab. `L` and `M` are interested in two companies: Almighty solutions Ltd. (A) and Brave computing Inc. (B), and they know that nobody else applied for them. The companies are not players in the game: Their decision is just a random process independent for each candidate. A company is more likely to hire better qualified applicants, which happens as follows. If a company only receives one application, they will accept it with 50% probability. If both `L` and `M` apply to the same company, the acceptance probability of `L` (who is more qualified) is 50% while `M`'s is 20%. However, these events are independently drawn, and the company may even hire them both (or neither of them). Both `L` and `M` have the same evaluations of the companies: Doing the internship at A is worth 100 while doing it at B gives utility 110, and being rejected and forced to choose the university lab gives 0. Additionally, if they manage to spend their internship together at the same company, their utility is doubled. Finally, `L` and `M` choose their application independently. They are rational players and what described above is common knowledge to them.

1. Represent this game in normal form.

2. What is the Nash equilibrium of this game in pure strategies?

3. Find all the Nash equilibria of the game.

Solution

1. This is a static game of complete information. As mentioned, the only players are `L` and `M`, since the behaviors of the companies can be modeled as random interventions by Nature, in other words, lotteries. Thus, the normal-form representation is a bi-matrix, where the set of players is $\{\mathtt{L}, \mathtt{M}\}$ and their common strategy set is $\mathcal{S}_{\mathtt{L}} = \mathcal{S}_{\mathtt{M}} = \{A, B\}$.

   The payoffs in the normal form are written by considering the *expected value* of the utilities. If they both choose the same company, neither of them is hired with probability 40%; only `L` is hired also with 40% probability; and finally, 10% is the value of both probabilities of (i) only `M` being accepted or (ii) both of them being accepted. This means that $u_L(A, A) = 40 + 20 = 60$, $u_M(A, A) = 10 + 20 = 30$. For (B,B) the utilities can be computed with the same approach, but are both multiplied times 1.1, since company B is ten percent better than A. If they choose different companies, the one going to A gets an expected utility of 50, while the other gets 55.

M

	A	B
L A	60, 30	50, 55
B	55, 50	66, 33

2. It can be seen that there are indeed no Nash equilibria in pure strategies. The best response of L is to copy M's move, while M does the opposite.

3. Then, there must be a Nash equilibrium in mixed strategies, which has both A and B in its support for both players.

   To find this Nash equilibrium, consider λ and μ as the probabilities that L and M play A, respectively, and exploit the indifference principle. For player L, it must be that $u_{\mathtt{L}}(A, \mu) = u_{\mathtt{L}}(B, \mu)$, i.e.

$$60\mu + 50(1-\mu) = 55\mu + 66(1-\mu),$$

   which leads to $\mu = 16/21$, while for player M the condition $u_{\mathtt{M}}(\lambda, A) = u_{\mathtt{M}}(\lambda, B)$ must hold, i.e.

$$30\lambda + 50(1-\lambda) = 55\lambda + 33(1-\lambda),$$

   which leads to $\lambda = 17/42$.

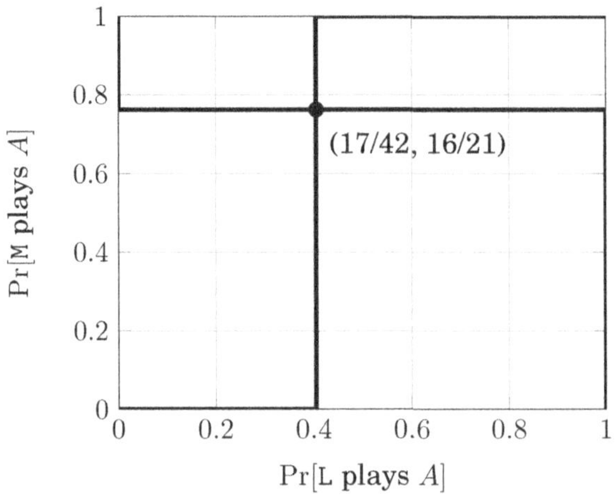

Figure 4.1: Mixed Nash equilibrium of the game.

Exercise 4.2. Arthur (A) and Crimilde (C) are rival hunters, looking for preys in the woods. They can choose to hunt:

- a bear (B),
- a deer (D),
- or a wolf (W).

Of these, they can only choose one, without consulting with each other. However, all the information given in the following is common knowledge for both of them. They know that if they select the same prey, they will make too much of a noise and the animal will flee. Thus, the payoffs will be 0 for both of them. Also, if each of them hunts a prey alone, the payoff of this solitary hunter will be $+1$ if successful in catching the animal and bringing it home. This will happen with probability ρ_j with $j \in \{B, D, W\}$, while with probability $1 - \rho_j$ the hunt goes wrong and the animal kills the hunter, in which case the payoff of the hunter is -1. These probability values are $\rho_B = 0.4$, $\rho_D = 0.9$, $\rho_W = 0.6$, and are also common knowledge.

1. What kind of game is this? Write down its normal form.
2. Find the Nash equilibria (NEs) in pure strategies.
3. If there are NEs in mixed strategies, find them. If not, discuss why.

Solution

1. Even though some actions seem to imply a dynamic evolution (such as, the hunter chooses a prey, and after this choice, they fight and either of them is killed), this is clearly a static game of complete information. In fact, all payoffs are computed immediately, based on the joint choices of the rational players A and C. In this, the behavior of the prey, which can be killed or turn the tables and kill the hunter, is just randomly drawn as a lottery.

   So this exercise is true to form in having humans as rational players and the animals as moved by Nature. Possibly in the future this will be considered speciesism, but at least for the purposes of this exercise, it looks justified.

   Thus, the normal form is as follows. The set of players is $\{\mathtt{A}, \mathtt{C}\}$. The strategy sets are $\mathcal{S}_{\mathtt{A}} = \mathcal{S}_{\mathtt{C}} = \{B, D, W\}$. To represent the payoffs, we need to resort to expected utilities, since some of the choices result in lotteries. Specifically, if both hunters choose the same action (i.e., pure strategy), they get a degenerate lottery where their certain payoffs are both 0. Otherwise, they actually hunt different preys and each get a separate lottery, where the payoff is $+1$ or -1 with probabilities ρ_j and $1 - \rho_j$, respectively.

   For example, a lone hunter chasing a deer ($j = D$) gets expected utility $+1(\rho_D) - 1(1 - \rho_D) = 0.8$, while for a bear the expected value is $+1(\rho_B) -$

$1(1 - \rho_B) = -0.2$. This reflects that a bear is less convenient to hunt than a deer.

The resulting bi-matrix is thus:

C

A	B	D	W
B	0, 0	-0.2, 0.8	-0.2, 0.2
D	0.8, -0.2	0, 0	0.8, 0.2
W	0.2, -0.2	0.2, 0.8	0, 0

2. The NEs in pure strategies are found to be (D, W) and (W, D). We can show that choice B is strictly dominated by both D and W.

3. There is also a mixed strategy equilibrium, where both players mix D and W. The mixture is found by setting indifference between the strategies in the support and correspondingly setting the same values to the expected payoffs. If α is the probability that a player chooses D, we have $0.8(1-\alpha) = 0.2\alpha$ that results in $\alpha = 4/5$.

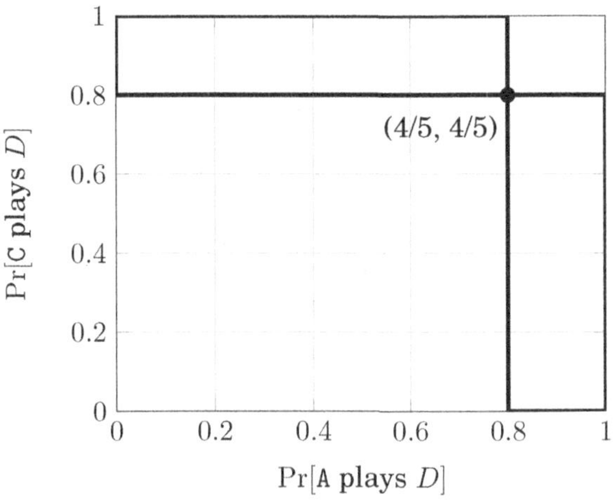

Figure 4.2: Mixed Nash equilibrium of the game.

Exercise 4.3. Jean (J) and Kate (K) want to buy a design handbag of new trendy brand, but it is difficult to find one in town. Only two places sell this handbag: a retail boutique R downtown, or the store S at the shopping mall. Both locations only have just one piece of the handbag, and J and K have time to go to only either of these places; moreover, if they go to the same place they have a $50/50$ chance of buying the handbag (it depends on who arrives first at the shop, which is entirely random). Their evaluation of the handbag is 600 for J and 450 for K. The official cost of the handbag is 250 euros, but shop R is known for giving a discount of 20% on all list prices. The utility for buying a handbag is evaluation minus cost; ending up without the handbag just gives utility 0.

1. Write down the normal form of this game.

2. Find all Nash equilibria of this game in pure strategies.

3. Find all Nash equilibria of this game in mixed strategies.

Solution

1. The set of players is {J, K}. Their strategy sets are $\mathcal{S}_{J} = \mathcal{S}_{K} = \{R, S\}$. When J and K choose the same pure strategy, their payoffs can be computed as expected utilities in a lottery. Actually, this results in them just getting half of the utility that they would get if they shopped alone.

   As a side note: Compare the result of the lottery with the discount on the price given by the retail boutique, which is instead an *actual* percentage and does not involve any lottery. Still, the involved computations are of the same kind, in the end. One just need to take into account the expected utility of uncertain events by taking the probabilities as weights.

   In other words, assume that two situations are compared where a customer can (i) get a handbag of given value V and paying a certain price $P < V$ with a 50% chance (and going empty-handed in the other 50% of the cases); or (ii) obtain with certainty a different handbag of lower quality, whose value is $V/2$ but also costs half the price, i.e., $P/2$. Following the vNM-utility theory, as commonly done in Game Theory, the outcomes of (i) and (ii) are entirely equivalent (or perfect substitutes, in economic terms).

   Thus, the normal form of the game can be represented by the following bi-matrix:

		K: R	K: S
J	R	200, 125	400, 200
	S	350, 250	175, 100

2. The game has 4 possible outcomes in pure strategies. The Nash equilibria in pure strategies are two: (R, S) and (S, R) (this is a discoordination game).

3. In addition to those, there is also a mixed strategy equilibrium. To find it, consider both players playing a mixed strategy and call a and b the probabilities that J and K play R, respectively (so they play S with probabilities $1-a$ and $1-b$). Both values are found by equating $u_{\mathtt{J}}(1,b) = u_{\mathtt{J}}(0,b)$, $u_{\mathtt{K}}(a,0) = u_{\mathtt{K}}(a,1)$, which leads to:

$$125a + 250(1-a) = 200a + 100(1-a) \text{ i.e, } a = 2/3, \tag{4.1}$$

$$200b + 400(1-b) = 350b + 175(1-b) \text{ i.e., } b = 3/5. \tag{4.2}$$

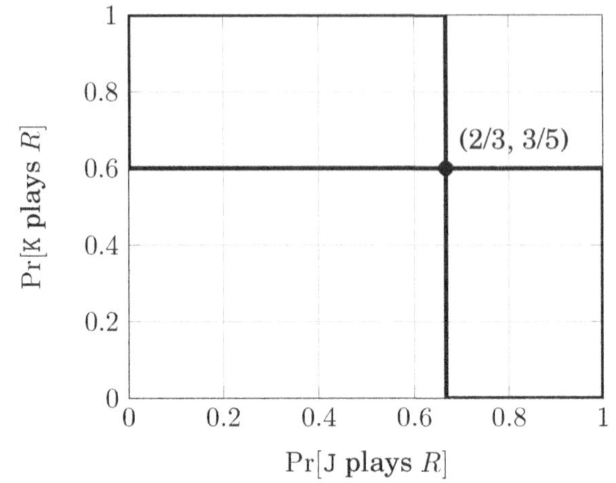

Figure 4.3: Mixed Nash equilibrium of the game.

Exercise 4.4. Anastasia (A) and Bruno (B) are friends enrolled in the same major and they need decide an optional course for the incoming semester. Their options are:

- Game theory (G)
- Hyperspectral imaging (H)
- Information theory (I)

Of these, they can only choose one, and they do so without consulting with each other. However, all the information about the courses given in the following is common knowledge for both of them.
They know that if they select a different course, they will not meet in classroom and this will make them sad. Thus, the payoffs will be 0 for both of them. Also, even if they meet in the same class, their payoffs will be non-zero only if the course is good. This will happen with probability ρ_j with $j \in \{G, H, I\}$, while with probability $1 - \rho_j$ the course is bad, in which case they will stop attending the lectures and in the end they both will also have payoff zero. These probability values are $\rho_G = 0.4$, $\rho_H = 0.8$, $\rho_I = 0.5$, and are also common knowledge.
In case the course is good, player A will have payoff $+1$ no matter what course it is. Player B will have payoff $+3$, $+2$, -1, for G, H, I, respectively (Bruno will realize afterwards that he does not like Information theory at all).

1. Write down the game in normal form. Find a suitable way to express the payoff of each outcome (explain it).
2. Find the Nash equilibria in pure strategies.
3. Are there Nash equilibria in mixed strategies? If yes, find them. If not, discuss why.

Solution

1. This is a static game, which admits normal form representation using a bi-matrix. The set of players is {A, B}, and the strategy sets are $\mathcal{S}_A = \mathcal{S}_B = \{G, H, I\}$. To represent the payoffs, we need to resort to expected utilities as follows:

B

A		G	H	I
	G	0.4, 1.2	0, 0	0, 0
	H	0, 0	0.8, 1.6	0, 0
	I	0, 0	0, 0	0.5, -0.5

2. The Nash Equilibria in pure strategies are found to be (G, G) and (H, H). We can show that choice I is strictly dominated for player B and therefore can also be iteratively eliminated as a further dominated strategy for player A. This IESDS process is in mixed strategies: G only weakly dominates I, and so does H, but a linear combination of G and H dominates I.

3. There is also a mixed strategy equilibrium, which is found by setting and equality between payoffs. If α and β are the respective probabilities that `A` and `B` play G, we have $1.2\alpha = 1.6(1-\alpha)$ that results in $\alpha = 4/7$, and $0.4\beta = 0.8(1-\beta)$ which means $\beta = 2/3$.

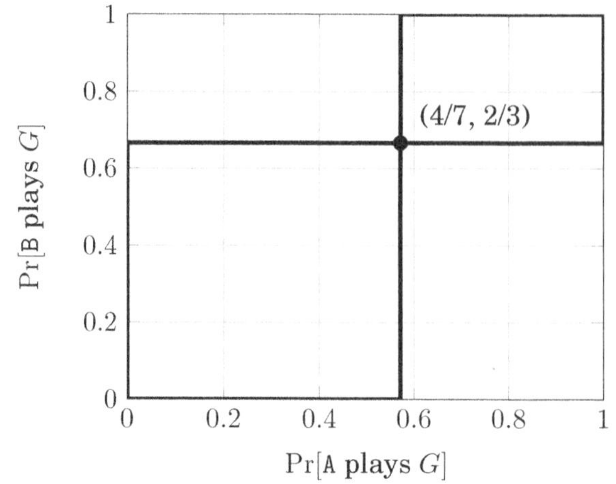

Figure 4.4: Mixed Nash equilibrium of the game.

Exercise 4.5. The Black Knight (B) and the White Knight (W) fight a duel to the last blood. They can choose (simultaneously and independently of each other) a weapon: Axe (A), Mace (M), or Sword (S). The knights are fully aware of the possible outcomes, which are as follows.

If they both choose the same weapon, the White Knight has probability 60% of winning. If either of them chooses Axe and the other chooses Mace, whoever chooses Axe has probability 90% of winning. If the Black Knight chooses Sword (and the White Knight does not), the Black Knight has probability 80% of winning. If the White Knight chooses Sword (and the Black Knight does not), the White Knight has probability 70% of winning.

Winning gives a utility of $+10$, dying has a utility of -100.

1. Write down the normal form of this game. Find a suitable way to express the payoff of each outcome (explain it).
2. Find the Nash equilibria in pure strategies.
3. Are there Nash equilibria in mixed strategies? If yes, find them. If not, discuss why.

Solution

1. The normal form is as follows: Set of players is $\{\mathtt{W}, \mathtt{B}\}$. The strategy sets are: $\mathcal{S}_{\mathtt{W}} = \mathcal{S}_{\mathtt{B}} = \{A, M, S\}$. To represent the payoffs, we need to resort to expected utilities. Also, note that the game is competitive (each outcome has a total payoff of -90, so we can sum 45 to each player's payoff and get a zero-sum game). The bi-matrix is:

B

W		A	M	S
	A	-34, -56	-1, -89	-78, -12
	M	-89, -1	-34, -56	-78, -12
	S	-23, -67	-23, -67	-34, -56

2. We can find the Nash equilibria by applying IESDS. We first eliminate M for both players, that is strictly dominated by S, obtaining the following reduced bi-matrix.

B

W		A	S
	A	-34, -56	-78, -12
	S	-23, -67	-34, -56

Now A is dominated by S as well and can hence be safely removed, leaving (S, S) as the only survivor of IESDS. Therefore, (S, S) is the only Nash equilibrium of the game.

3. Since there is a unique Nash equilibrium in pure strategies, there are no other equilibria in mixed strategies.

Exercise 4.6. The Count of Carmanova (C) is defied by the Duke of Dondestan (D) to a duel. Thus, the Count can choose the weapon, either sword (S) or pistol (P), and the Duke can reply with a choice on whether to duel to the first (F) or last (L) blood. If S is chosen, the Count wins 40% of the times (and the Duke 60%). If P is chosen, the Count wins 70% of the times, the Duke 20% of the times, and 10% of the times they both get fatally shot. If the duel is to the last blood, the loser always dies. Otherwise, the loser survives (unless it is a duel with pistols and both assailants get fatally shot), but he is dishonored. For both players, dying gets payoff -100. Surviving but being dishonored corresponds a payoff of -20. Winning the duel gives payoff $+10$.

1. Represent this game with C and D in extensive form.
2. Represent this game in normal form by expanding the strategies of D, and find the Nash equilibria.
3. Discuss which Nash equilibria are subgame perfect.

Solution

1. The extensive form is the tree depicted in Figure 4.5. To get the numbers right, one must compute the expected payoffs of any duel.

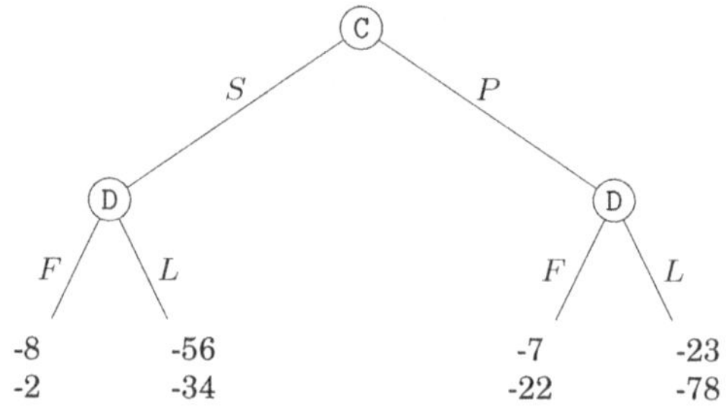

Figure 4.5: Extensive form of the game.

2. A normal form representation is

W \ M	FF	FL	LF	LL
S	-8, -8	-8, -8	-56, -40	-56, -40
P	-7, -24	-23, -80	-7, -24	-23, -80

There are three Nash equilibria: (S, FL), (P, FF) and (P,LF).

3. However, when searching for subgame perfect equilibria, we notice that some of these Nash equilibria involve non-credible threats. In particular, (S,FL) basically means that the Duke threats the Count to fight to the

last blood if pistol is chosen, and concedes first blood if sword is chosen. This makes sense for the Duke, who is better with the sword than with the pistol. Still, it is a non-credible threat, since the Duke is afraid to die as much as the Count, so F is always preferable to L for him. Instead, (P,LF) is identical to (P,FF) as the outcome is anyways PF, but the former has a non-credible behavior off-equilibrium. Thus, the only SPE is (P,FF).

Exercise 4.7. David (D) and Emily (E) are baking a cake, according to the following plan. David will choose a type of cake, Emily will buy the ingredients (being aware of David's choice), and they will bake the cake together. For the first step, David knows three types of cakes: Apple pie (A), Brownie (B), or Cheesecake (C). Denote his choice by $i = \{A, B, C\}$. For the second step, Emily can buy the ingredients at either the Farmer's market (F) or the Grocery store (G). This applies to any cake that David chooses in the first step. Denote her choice as $j = \{F, G\}$. The last step is actually played with a random intervention by Nature, in that they have a successful baking with probability $p_{i,j}$ that is a function of i and j as reported in the following table.

D \ E	F	G
A	0.5	0.2
B	0.8	0.4
C	0.7	0.9

All these values are common knowledge to David and Emily. Emily does not have any preference between the cakes, her utility u_E is 1 if the cake is successfully baked, 0 otherwise, regardless of its type. David instead has a preference for some cakes, so his utility u_D is equal to: $3u_E$ if the cake is A; $2u_E$ if the cake is B; u_E if the cake is C. Note that even David still values the utility of an unsuccessful cake to be 0 regardless of the type.

1. Represent this game in extensive form.
2. Represent a "flat" version of this game (as if all moves were simultaneous) in normal form.
3. Find the Nash equilibria of the game and discuss which are subgame-perfect.

Solution

1. The extensive form is the tree depicted in Figure 4.6. D chooses first among A, B, and C. Then E chooses either F or G. This results in six possible combinations.

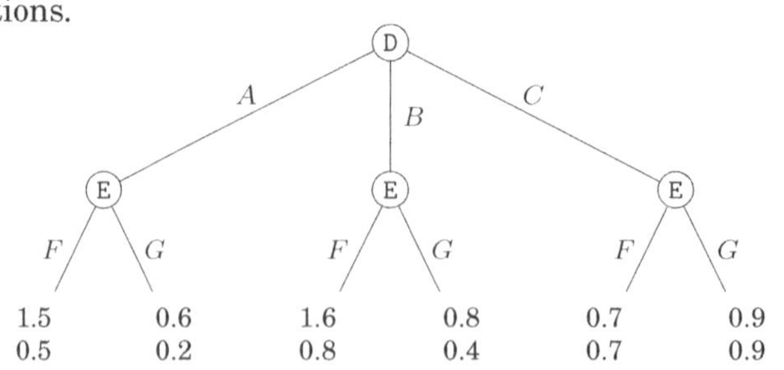

Figure 4.6: Extensive form of the game.

2. A normal form representation of the corresponding "flat" game is as follows.

D \ E	F	G
A	1.5, 0.5	0.6, 0.2
B	1.6, 0.8	0.8, 0.4
C	0.7, 0.7	0.9, 0.9

A is a strictly dominated strategy for D and can safely be removed. We can hence consider a reduced version of the game in which the available strategies for D are $\mathcal{S}_D = \{B, C\}$.

3. Since E moves after D, her strategies must consider both the case in which D plays B and that in which he plays C. For example, strategy FG means "play F if B is played, G if C is played" (again, we do not consider A, as it is strictly dominated).

D \ E	FF	FG	GF	GG
B	1.6, 0.8	1.6, 0.8	0.8, 0.4	0.8, 0.4
C	0.7, 0.7	0.9, 0.9	0.7, 0.7	0.9, 0.9

There are three Nash equilibria: (B, FF), (B, FG) and (C,GG). Among those, (B, FG) is the only one which is also subgame-perfect, as one can verify by applying backward induction.

Exercise 4.8. A class of 13 students needs to select the next class rep for the academic year. The three candidates are Alicia Allen (**A**), Bill Baker (**B**), and Connie Carter (**C**). The principal was clear on allowing only one representative per class, which is chosen as the candidate with most votes. Using symbol $\succ$ to denote "preferred to", the common knowledge is that students' preferences are divided as follows:

- for students 1–5 **A**$\succ$**B**$\succ$**C**;
- for students 6–9 **B**$\succ$**C**$\succ$**A**;
- for students 10–12 **C**$\succ$**B**$\succ$**A**;
- for students 13 **C**$\succ$**A**$\succ$**B**.

1. What is the outcome of the election under *sincere voting*? Who is the winner W under this assumption?
2. How can the third group of voters (10–12) tamper with this result with a *strategic voting*? Who becomes the winner W'?
3. If this group applies strategic voting, do the supporter of W have some countermeasure to this? What is the reason for that?

Solution

1. This election is a simple majority voting, where the candidate with most preferences wins. Under sincere voting **A** gets preferences from students 1–5 (5 votes), **B** from 6–9 (4 votes), and **C** from 10–12 and from 13 (4 votes). Therefore, **A** wins the election.
2. For group 10–12, **B** is a preferable candidate to **A**. Since the group can foresee **A**'s victory, a strategy that they can adopt is to vote for **B**. This way, **B** is the new winner.
3. The only thing that students 1–5 could do to prevent **B**'s victory is to vote for **C**, which would result in a tie. However, this goes against their own interest, as they prefer **B** to **C**. Thus, there is no countermeasure that they can adopt. Notice that **B** is the Condorcet's winner so it is difficult to tamper against him.

Exercise 4.9. The mayor of a big city is to be selected among four candidates: Amy Adams (**A**), Brandon Butler (**B**), Cornelius Coleman (**C**), and David Douglas (**D**). Using symbol $\succ$ to denote "preferred to", polls indicate that:

- **A** has 42% of supporters. Also, for them **B**$\succ$**C**$\succ$**D**.
- **B** has 11% of supporters. Also, for them **A**$\succ$**C**$\succ$**D**.
- **C** has 27% of supporters. Also, for them **B**$\succ$**D**$\succ$**A**.
- **D** has 20% of supporters. Also, for them **C**$\succ$**B**$\succ$**A**.

1. The election is being held as a two-round run-off (i.e., with a ballot). What is the outcome under sincere voting? Denote the winner as W.
2. Assume that the supporters of **D** can identify this outcome and plan a strategy. What is the best strategic voting that they can enact?
3. Discuss the identity of the winner W' under strategic vote of **D**'s supporters. What kind of choice is W'? Can the supporters of W prevent this outcome by counteracting strategic vote of **D**'s supporters, with a strategic vote of their own?

Solution

1. Under sincere voting, **A** and **C** are the candidates with most supporters, so they go to the ballot. At the ballot, **A** gets also the votes of **B**'s supporters – who prefer **A** over **C** – reaching a total of 53%. This makes **A** the winner of the ballot (and of the election).
2. **D**'s supporters prefer **B** over **A**. Thus, they can support (**B**) in the first round, so that the candidates going to the ballot are **A** and **B**. Now at the ballot **B** gets the support of 58% (everybody but **A**'s supporters).
3. The new winner under strategic vote of **D**'s supporters is indeed **B**. Despite having the lowest number of first-choice votes, **B** is the Condorcet winner, thus he surely wins in a one-to-one match. **A**'s supporters cannot avoid the victory of **B** if he goes to the ballot: their only alternative is to keep sending **C** to the ballot instead, Thus, they should split, e.g., with $7/42$ of them supporting **C** instead of **A** at the first round. This way, at the first round **A** gets 35%, **B** gets 31%, **C** gets 34%. **A** and **C** go to the ballot, and **A** wins.

Exercise 4.10. Marcel (M) and Noah (N) are playing a simultaneous game, where the pure strategies available to M are A, B, and C, whereas N can play X, Y, or Z. The payoffs are represented by the following bi-matrix when put in normal-form.

N

M		X	Y	Z
	A	2, 2	3, 1	0, 0
	B	1, 6	7, 4	9, 4
	C	0, 1	5, 3	9, 2

1. Find all the Nash equilibria (NEs).
2. Consider a Stackelberg version of this game where M is leader and discuss it.
3. Consider a Stackelberg version of this game where N is leader and discuss it. In this specific case, if you need a rule to break ties, you might assume that the leader and the follower are generous.

Solution

1. To find the NEs, we start by highlighting the best responses. M's best responses are overlined, N's best responses are underlined.

N

M		X	Y	Z
	A	$\underline{\overline{2, 2}}$	3, 1	0, 0
	B	$\underline{1, 6}$	$\overline{7, 4}$	$\overline{9, 4}$
	C	0, 1	$\underline{5, 3}$	$\overline{9, 2}$

It is found that the only NE in pure strategies is (A,X), with payoffs equal to 2 for both players. A quick search on mixed strategies reveals that this is also the only NE of the entire game.

2. The Stackelberg version of this game with M as the leader is actually a dynamic game. This can be solved through backward induction, and it is simple to find the solution (sometimes called the Stackelberg equilibrium) even without drawing the extensive form. Indeed, rationality of the players implies that, whatever M chooses, N will play a best response to it. But then, M can exploit his first mover advantage and anticipate that N will do so. In essence, M is comparing the three underlined cells in the previous table and considers the one giving him the best result, which turns out to be (C,Y).

   We note two additional properties. First, M is guaranteed not to go below the payoff at the NE (A,X) of the static version. This is because (A,X) must be one of the compared cells, so M can force this result to be achieved in the Stackelberg version. But in this case he even strictly improves over it, because he has the better alternative (C,Y) available. Also N is guaranteed to get at least his minimax, since by definition the minimax is the minimum payoff that a player gets when moving with knowledge of the other player's choice. This is exactly what happens to `N` here, since he is the follower. In reality, `N` strictly improves over his minimax, since he gets 3 at the Stackelberg equilibrium and his minimax was 2.

These improvements (of `M` over the static NE and of `N` over the minimax) are not surprising, since the game is *not* competitive. Thus, translating the simultaneous move into a Stackelberg game actually improves the coordination of the players. The leader `M` is guaranteed to get an improvement because he has first mover advantage, but this does not mean that `N` must be worse off, he is actually improving his payoff too, because of the existence of a better outcome in Pareto sense.

In practice, one can even consider this applied in a practical context. The difference between the standard static game and the Stackelberg version justifies whether one of the players should act as the leader and makes its decision public to the other. This generally has a communication cost in real cases, and besides, the other player is keen to listen to this only if being the follower is advantageous, which is not always the case, and is especially false if the game is competitive.

3. The game version with `N` leader follows the same reasonings than the previous case, but with an additional catch. Since solving the Stackelberg game essentially boils down to apply backward induction, we note that in this game some payoffs are identical in value, which may be a problem for getting a clear solution from backward induction. We actually need some tie-breakers and the text specifies that both leaders and followers are generous. What does it mean and why does it lead to a tie-breaking rule here?

   Observe that `N` must now compare the overlined cells, i.e., the best responses chosen by `M`. But actually, `M` has *two* best responses to Z (both B and C) so it is unclear which one to choose. The principle of "generous follower" implies that `M` will choose what is better for the leader, which in this case is (B, Z).

   Still, we now face another tie, because `N` must now choose the best outcome among (A, X). (B, Y), and (B, Z). Of these, both (B, Y) and (B, Z) are better than the former NE of the static game (A, X), but they are absolutely equivalent for `N`. To solve the indeterminacy, we apply the "generous leader" principle and assume that also `N` returns the favor and chooses the outcome that is better for `M`.

   As a side note, these assumptions of generosity only make sense exactly because, as remarked before, the game is non-competitive. In a zero-sum game, they would not make sense but they will not be useful either, since a tie for the leader implies that also the follower gets the same payoff and vice versa, as the two players' payoffs are always opposite to each other.

5 Multistage Games

Multi-stage games are a class of *dynamic games* where smaller games are played over multiple subsequent *stages* and combined into a bigger global game. In particular, the general theory of multi-stage games is especially applied to repeated games, in which all stages are alike, which also allows for infinite-horizon with interesting conclusions.

5.1 Characteristics

Stage games are often chosen as simple 2-person static games of complete information. They can possibly be a modified version of well known games such as the prisoner's dilemma, the coordination game, or the discoordination game. They try to capture the strategic impact of current actions of some players to the future actions of other players.

The resulting multistage game is the combination of these individual smaller games, assuming that they are always played by the same agents, and that the payoffs of the stage games are aggregated in some way. The most common approach is to consider that discounted versions of the local payoffs are cumulated to get the total payoff in the grand game. The discount factor δ is a value between 0 and 1 that reflects how much players care about the future. In economics, this is usually connected to resource decay or inflation of prices. But another appealing interpretation assumes that players can also abandon the game earlier than the envisioned time horizon and connects δ with the expected time spent in the system (the lower δ, the earlier the player leaves the game).

Some problems consider a multistage version made of different games combined together. For the sake of simplicity, these cases are usually limited to the aggregate of two different stage games. However, it is also very common to consider *repeated* games, in which the same stage game is played multiple times, and in this case it is common to see the game repeated for more than 2 times. Actually, a very interesting case is that of the game being repeated with an infinite horizon. While this is impractical to consider for the aggregate of different games, it is definitely the most common case for repeated game scenarios, especially if the underlying model is not that the players are actually doomed to play a game for all eternity, but actually they do not know the last stage of their interaction (see again the connection between δ and their probability of hanging out in the system for the next round).

5.2 Finite horizon

The solution approach for finite horizon is analogous to backward induction. One must consider the *last* stage, since in that case there is no impact on future decisions, which are non-existent, and therefore the players simply choose an NE (i.e., they very selfishly act like "there is no tomorrow"). Then, this conclusion is propagated back to the penultimate stage and so on.

This leads to different conclusions depending on the numbers of NEs in the stage games. If the last stage game has a single NE, then the rational players are fully aware of that and they all know that this will be what they are going to play in the end. Thus, the last stage can be safely removed from the analysis by "merging" it with the upper layers and the game just become shorter.

In the particular case that *all* stage games have a single NE, as is the case for a repeated version of a static game with a single NE, the conclusion is immediate, but not very interesting. The only rational outcome is that this single NE is played at every stage, and there is no strategic extension in the unfolding over time, beyond what already given by a one-shot game. Notably, the repetition of the NE at every stage will also lead to an SPE, since an NE is played in every subgame. But just keep in mind that an SPE will require some more details to be specified, such as what happens off the equilibrium path – this is immediate to determine, since what happens is that the only NE of the stage game is played there too, it is just that it is easy to forget to mention it.

If there are multiple NEs available instead, some strategic variants become available, and this implies that some non-NE outcomes of the stage game can be played. Naturally, this cannot happen for the last stage, which is bound to be played as an NE, but the choice among multiple NEs in the last stage may even create some "carrot and stick" options where players create a collaborative outcome in the earlier rounds. This happens by defining one good NE of the last round as the ultimate reward (carrot) and a bad NE of the last round as a punishment (stick) and anticipating that cooperation in the earlier stages will be rewarded, while defection will be punished.

For this approach to work, some conditions are required. First, the difference between reward and punishment must be high enough. And at the same time, the discount factor, if present, ought to be high enough to make the punishment credible. This is particularly evident if δ is low since the players will care little about the future; in particular, if the interpretation of a low δ is that they are likely to exit the system, they will consider this as their last stage, so there is no way to convince them to play something else than an NE.

Overall, the conditions for this approach to work are once again connected with the concept of credible threats. Since there is actually no communication and the interaction of punishment and/or reward only happens in the players' heads, they can anticipate that a possible behavior is to get punished if they do not cooperate, but if this threat is non credible, they will dismiss it, so the dynamic outcome will not result in an SPE. Thus, the request to find an SPE in such cases translates into looking on whether these conditions support such a reward/punishment approach, which usually translates into a requirement on δ that must be above a certain value. Keep in mind that the standard case of playing an NE multiple times in every evolution of the game is also always present as an SPE.

5.3 Infinitely repeated games

For the case of games that are repeated infinitely many times, the conclusion is even more interesting. Instead of requiring an explicit difference between two or more NEs to have a credible reward and punishment approach, this can be sort of created out of thin air by the so called "grim trigger" approach. This is true even for repeated games with a single NE, and indeed it is often studied in application to the prisoner's dilemma. Whatever the underlying stage game, it is usually assumed that the players have a cooperating and a defecting option, which in the case of the prisoner's dilemma are to collaborate with police or stay silent, respectively. Only, cooperation is not an NE of the stage game, while mutual defection is.

In an iterated game that repeats such a stage game infinitely many time, it is found that playing the NE forever is of course a possible SPE. But also cooperation played forever can be established as an alternative SPE, or better, as its equilibrium path. To prove this, one cannot resort to an approach akin to backward induction, since there is actually no last stage to start from. The idea is that the players start with an initial cooperation and they "agree" (once again, this agreement is just in their head since they do not actually communicate) that deviating players will be punished for the rest of the game, which is forever. This is formally defined as a "grim trigger" strategy, which is to play cooperation as long as all previous stage results describe mutual cooperation, and to defect otherwise. It is called like that because once the trigger is pulled, there is no way back.

This strategy ought to establish cooperation being played forever as the equilibrium path. Yet, it does not fully prevent a player from deviating since selfish players are naturally tempted to adopt a myopic deviation, i.e., to play a deviation in the stage game. Such a deviation must exist and is actually the reason why the cooperating option is not a local Nash equilibrium in the first place. However, if this player also believes that the other player is adopting the grim trigger strategy, then the player anticipates that the myopic deviation will be advantageous for the current stage but will also lead to reduced payoffs in the future. Thus, the grim trigger is effective as long as the utility of cooperating forever is higher than taking a selfish deviation now and facing an eternal punishment in the future.

Notably, this depends as before on the difference between reward and punishment and also the value of the discount factor. If δ is low, even the threat of an eternal punishment may seem not credible – which may also well apply to ethical conclusions under a religious assumption. From a more mundane perspective of a game theoretic exercise, this means that very often the request boils down to computing the minimum δ that establishes cooperation, which implies that the discount factor must fall into an interval like $(\delta_{\min}, 1)$ to have cooperation.

Beyond the standard case of the simple choice between cooperation and defection, more complex cases can be created whenever there are worse punishments available than eternal defection. Indeed, eternal defection may seem already scary enough, but there may be something like an immediate punishment *and* also eternal defection. This leads to an even lower required δ to establish cooperation. To better understand why, consider the following example that once again digresses into ethics. A social cooperation rule is generally established by threatening punishment. The more immediate, or the stronger, the punishment, the better the cooperation. If you want a society to behave well, for example, not cause global warming through environmental pollution, you can claim that this will lead to the disruption of the planet forever. Unfortunately, this eternal destruction of the planet does not seem very menacing for boomers whose life expectancy is low, and therefore they do not care much if they are going to leave a polluted Earth to their grandchildren. So, establishing an extra punishment in addition to that, like sending them to jail or causing them to pay a lot of money may be more effective.

The reasonings above can actually be extended also in the direction of not only obtaining just cooperation but a more involuted outcome thanks to *Friedman's theorem.* Simply put, the theorem states that a scenario where players interact for infinitely many rounds can be driven to any linear combination of strategies that satisfies a certain condition if the discount factor is high enough. The condition is naturally that the resulting payoffs of all players must be better in a Pareto sense (i.e., element-wise) than the payoffs at the NE of the stage game. Clearly, it is impossible to convince a player to stick to a strategy that awards an eternally worse payoff than the one at the NE: at this point, the player would deviate and play the NE forever. However, any linear combination of moves (even over subsequent multiple rounds) that leads to an expected payoff higher than the NE for all players is sustainable as long as δ is high enough. The theorem essentially generalizes the reasoning of the grim trigger mentioned above to the case where the objective is not to play the basic cooperation, but possibly a mixed strategy and/or different strategies over multiple rounds.

Exercises

Exercise 5.1. A guitarist (G) and a pianist (P) are playing together two songs (denoted as songs 1 and 2) on stage for a jam session. For both songs, they can decide to sing (S) or just play (J). They decide this independently and without consulting each other, for each song. If both of them stay silent (not singing, just playing) a song, the song will not be great and the payoff for that song will be 0 for both. If only one of them sings, the song rendition is better. However, they prefer to be the non-singing musician as they can concentrate on their solo. Thus, for that stage, whoever sings, gets utility 1, the one staying silent gets: 7 for the guitarist, and 6 for the pianist. If they both sing the song (as a duet), the payoff is 5 for both. The total payoff at the end is the sum of the two stages.

1. Consider the stage game: write it in normal form and find its (pure strategies) Nash equilibria.
2. Find out the subgame-perfect equilibrium where they play the same gameplay at every stage and that gives the total highest utility overall.
3. Is it possible to identify a subgame-perfect equilibrium where the first song plays out as a duet (they both sing)? If so, find it. If not, prove why.

Solution

1. The game is a repeated game with no discount, i.e. $\delta = 1$, which lasts for two stages. The normal form of the stage game is:

   P

G	S	J
S	5, 5	1, 6
J	7, 1	0, 0

   There are two Nash equilibria in pure strategies: (S,J) and (J,S), making the stage game an anti-coordination game.

2. Trivial subgame-perfect equilibria in the two-stage game are found by repeating a Nash equilibrium twice. Thus, for example, they play (S,J) twice. However, this gives a total of 14, the best one is to play (J,S) twice, which results in a grand total of 16.

3. The guitarist can also convince the pianist that he will play this strategy as a Grim Trigger: "Start by playing S. Then in the second round, I play S (and you can play J) only if the first outcome was (S,S). Otherwise, I play J instead." If this strategy is adopted by G, it is the best response for P to comply and play S in the first round, and J in the second (total utility of 11) as opposed to betraying in the first iteration (thus getting 6 but then getting 1 in the second stage game).

Exercise 5.2. Ashley (A) and Brook (B) live together. During the winter break they contemplate giving each other a nice gift (G) for Christmas – or not (N), because indeed presents are expensive to buy. They make this decision independently and without telling each other. After Christmas, they also consider whether to celebrate New Year's eve downtown (D) or stay home (H). They know each other's preferences so they are able to buy a gift for 10 euros that is worth like 100 euros for the other. This means that receiving a gift gives utility of 100 to the receiver that is moved by the gesture, but also -10 to the utility of the buyer. Not giving gifts implies no variation of the utility for both.

For the New Year's eve celebration, they decide independently of each other in a coordination-game fashion. Staying home has utility of 0 for both. Going downtown has utility of 50. However, spending New Year's eve apart from each other has utility of -100 for both. The total payoff of the players is the sum of the partial payoffs in each stage (the gift exchange at Christmas, and the party at New Year's eve) with a discount factor of δ for the second stage.

1. Write down the normal form of both stages of the multi-stage game.
2. Find a trivial subgame-perfect equilibrium of the game where the players just play a Nash equilibrium in all stages, without any strategic connection.
3. Find out if there is a strategically connected SPE of the whole game where Ashley and Brook give gifts to each other, and show the minimum required discount factor value $\delta_{\min}$ for that to hold.

Solution

1. In both stages, the set of players is {A, B}. The strategy sets are identical for both players and are: $\{G, N\}$ for the first stage, $\{D, H\}$ for the second.

   The normal-form representation is shown as the following bi-matrix for stage 1:

		B: G	B: N
A	G	90, 90	-10, 100
	N	100, -10	0, 0

   while for stage 2 it is this:

		B: D	B: H
A	D	50, 50	-100, -100
	H	-100, -100	0, 0

2. One trivial Nash equilibrium of the multi-stage game is for them to play Nash equilibria of the stage games, in sequence. Thus, both playing (N,D,D,D,D) or (N,H,H,H,H) are valid answers, as stage 1 has the only Nash equilibrium (N,N) and stage 2 has two, (D,D) and (H,H).

3. To support (G,G) in the first round, the players must play a stick-and-carrot NE where the choice is between the two Nash equilibria of the last stage. This means that they both play (G,D,H,H,H). This can be supported only if this strategy is the best response to itself. If A plays it, B's payoff when choosing it as well is 90 in the first round and 50 in the second one, for a total payoff of $90 + 50\delta$. A myopic deviation would be to play N in the first round, which results in payoff 100. This is not convenient if $50\delta > 10$ which means $\delta_{\min} = 1/5$.

Exercise 5.3. Ann (A) and Brian (B) moved in together, and they soon forget the excitement of going out and watch movies, now they play a new game called "who does the chores at home?" (cleaning, cooking, ironing, etc.). This game is repeated every single day, and plays out as follows. Both players can choose to do the chores (C) or not (N). If they both choose N, the utility for that day is 0 for both. If they both choose C, the house is very clean but they are very tired, so their utility of the day is -1 for both. If either of them does the chores whereas the others does not, the cleaner gets utility -2 while the lazy player gets the benefit of a clean house without any work, which is the best possible result and gives utility 5 for that day. The total utilities are aggregates of the individual day's utilities with a discount factor of 0.9.

1. Write down the stage game in normal form and find its Nash equilibria.
2. Prove (with a theoretical result) that a situation where A always does the chores and B does nothing is not sustainable for the infinitely repeated game.
3. Prove that a situation where A does the chores on even days and B does the chores on odd days is sustainable for the infinitely repeated game and write a Grim Trigger strategy to enable it.

Solution

1. The set of players is $\{A, B\}$. The strategy sets are $\mathcal{S}_A = \mathcal{S}_B = \{C, N\}$. The normal form is represented by the bi-matrix:

		B	
		C	N
A	C	-1, -1	-2, 5
	N	5, -2	0, 0

   It is evident that N is dominant and (N,N) is a Nash equilibrium of the stage game. Hence, a joint strategy of both players to "Always play N" at every stage is a trivial Nash equilibrium of the repeated game.

2. According to Friedman's theorem, we can sustain any feasible payoff (u_A, u_B) where $u_A, u_B > 0$ since the Nash equilibrium of the stage game gives $0, 0$ as outcome. The situation where A always does the chores gives her a negative payoff, though, and thus it is not sustainable.

3. The situation where the players alternate is sustainable, instead. A Grim Trigger strategy can be as follows. "At stage 1, B plays C and A plays N. At stage 2, B plays N and A plays C. At stages $t > 2$, with t odd, if the rule of alternating was kept, B plays C and A plays N. Otherwise, all play N forever after. At stages $t > 2$, with t even, if the rule of alternating was kept, B plays N and A plays C. Otherwise, all play N forever after."

This is sustainable, according to Friedman's theorem, since the expected payoffs are:

$$
\begin{aligned}
u_{\mathtt{A}} &= 5 - 2\delta + 5\delta^2 - 2\delta^3 \cdots = 5\frac{1}{1-\delta^2} - 2\frac{\delta}{1-\delta^2} \\
&= 5 \cdot 1.2345 - 2 \cdot 0.9 \cdot 1.2345 = 3.95 > 0
\end{aligned}
$$

for A and

$$
\begin{aligned}
u_{\mathtt{B}} &= -2 + 5\delta - 2\delta^2 + 5\delta^3 \cdots = -2\frac{1}{1-\delta^2} + 5\frac{\delta}{1-\delta^2} \\
&= -2 \cdot 1.2345 + 5 \cdot 0.9 \cdot 1.2345 = 3.086 > 0
\end{aligned}
$$

for B.

Exercise 5.4. Carl (C) and Diana (D) are two university students that have found that the department library is unoccupied overnight. It is a really good place to study and has a very fast Internet connection. So, they go there every night, but they do not coordinate or plan any action together. Upon their arrival every night, they independently decide whether: (S) study or (M) watch some movies on their laptop. If they both study, they both get utility 10. The individual benefit from watching a movie is instead 15 for C and 18 for D. However, if they both choose M, their individual benefit is halved (since they have half the connection speed). Also, trying studying while somebody else is playing a movie breaks the concentration, so $u_C(S, M) = u_D(M, S) = 0$ (C is written as the first player). Call $\mathbb{G}$ this game, and consider it in a repeated version $\mathbb{G}(T)$, where $\mathbb{G}$ is played every night for T nights. Individual payoffs are cumulated with discount factor δ. Finally, consider an *extended* game where a punishment strategy P is also available to both players. When either player chooses P, payoffs are -10 for *both* players (that would correspond, e.g., to do something really stupid in the library and get the library permanently closed). Call this game $\mathbb{G}'$. Note: despite P being weakly dominated, (P,P) is an NE for $\mathbb{G}'$.

1. Find the Nash equilibria of $\mathbb{G}(3)$, for $\delta = 1$.

2. What values of δ allow for sustaining a Nash equilibrium of $\mathbb{G}(\infty)$ via a "Grim Trigger" strategy where each player ends up in always choosing S?

3. If you see an SPE of $\mathbb{G}'(2)$ where players may play S, state at which round do they play it, and what value of δ do you need to obtain it.

Solution

1. The normal form matrix of $\mathbb{G}$ is the following.

		D	
		S	M
C	S	10, 10	0, 18
	M	15, 0	7.5, 9

   The only Nash equilibrium of the stage game is (M,M). This is also the only Nash equilibrium for a finitely repeated game.

2. The Grim Trigger strategy is: "start playing S at round 1, then keep playing it in round T as long as both played S in every round from 1 till $T-1$." This is a Nash equilibrium as long as deviating once and get (M,M) forever after is less advantageous. This requirement is more stringent for Diana. The payoff for both Carl and Diana playing S at every stage is

$$10 + 10\delta + 10\delta^2 + \cdots = 10\sum_{t=0}^{\infty} \delta^t = \frac{10}{1-\delta}$$

while the utility given by a deviation to M, for Diana, is

$$18 + 9\delta + 9\delta^2 + \cdots = 18 + 9\sum_{t=1}^{\infty} \delta^t = 18 + \frac{9\delta}{1-\delta}.$$

Solving $\frac{10}{1-\delta} \geq 18 + \frac{9\delta}{1-\delta}$ with respect to δ, leads to a constraint for the discount factor of $\delta \geq \frac{8}{9}$.

3. The normal form matrix of $\mathbb{G}'$ is the following.

D

C		S	M	P
	S	10, 10	0, 18	-10, -10
	M	15, 0	7.5, 9	-10, -10
	P	-10, -10	-10, -10	-10, -10

Game $\mathbb{G}'$ has two Nash equilibria, therefore we can apply a carrot-and-stick rationale, where a "carrot" is the better Nash equilibrium (M,M) and the "stick" is (P,P). The idea is to reward collaboration with the "carrot," and punish betrayal with the "stick." Formally, this translates into strategy $\mathcal{S}$ with a reward plan, defined as (i) start by playing S in the first round and (ii) in the second round, play M if the first round resulted in (M, M), otherwise (with every other result of the first stage) play (P, P). Note that (S,S) can only be played at the first round since at the last round the players are forced to play an NE of the stage game.

The idea would be that $(\mathcal{S}, \mathcal{S})$ is chosen as the NE of the whole game, and this ends up in (S,S) at the first round followed by (M,M) at the second. This actually happens if $\mathcal{S}$ played by one player is the best response to $\mathcal{S}$ played by the other, so we must avoid that deviations from $\mathcal{S}$ are advantageous. Actually, the only deviations that make sense can happen at the first round, since the last round is already played as an NE of the stage game, so there is no desire for unilateral deviation there. Still, we need to guarantee that $u(S, S) + \delta u(M, M) > u(\text{deviation}) + \delta u(P, P)$, because one player may consider deviating in the first round and then getting the "stick" P,P in the second, which can be still appealing if the deviation is convenient and the punishment is too weak. The strength of the punishment depends on the difference between "carrot" and "stick" but also on its certainty, i.e., the discount factor δ.

Deviating from $\mathcal{S}$ is more tempting for D, since she has the highest convenience in deviating, she will get utility 18 in the first round as opposed to 10. Still strategy $\mathcal{S}$ works if $10 + 9\delta > 18 - 10\delta$, i.e., $\delta > 8/19$.

Exercise 5.5. A warrior (W) and a mage (M) are exploring a dungeon looking for monsters. Their exploration is framed as a two-stage game. In the first stage, they prepare to fight a monster in the next room, and both of them can do either of two actions: cast a spell (C) or draw the sword (D); actually, strong W does not like to C and lazy M prefers not to choose D. Still, they choose their actions simultaneously without talking or communicating to each other in advance. Both players are fully aware of each other's partial payoffs for this stage, that are defined as follows: they get utility 2 if they choose different actions, while their utility is 0 if the actions are the same because this might make them less prepared against certain enemies. Moreover, 1 must be subtracted from the utility of a player who chooses the less-favorite action.

After that stage, a second stage is played where the two adventurers, while grabbing their loot, are surprised by a troll. Now, they can either run cowardly (R) or attack the troll (A). Also this action is decided indepedently of each other and at the same time. If they both attack, they successfully slay the troll (payoff 5 for both). A player attacking alone gets killed by the troll (payoff -100) while the other is unharmed and can easily take the treasure before running away (payoff 10). If they both run, they manage to escape but they must leave the treasure behind (payoff 0).

1. Draw the extensive form of this game.
2. Find all the Nash equilibria of the two-stage game.
3. Find all the *subgame-perfect* equilibria of the two-stage game and explain why are they subgame-perfect.

Solution

1. Before drawing the extensive form, it is convenient to put both stages in normal form. For the first stage, this is

W \ M	C	D
C	-1, 0	1, 1
D	2, 2	0, -1

while the normal form of the second stage is

W \ M	A	R
A	5, 5	-100, 10
R	10, -100	0, 0

The extensive form of the whole game is the tree depicted in Figure 5.1.

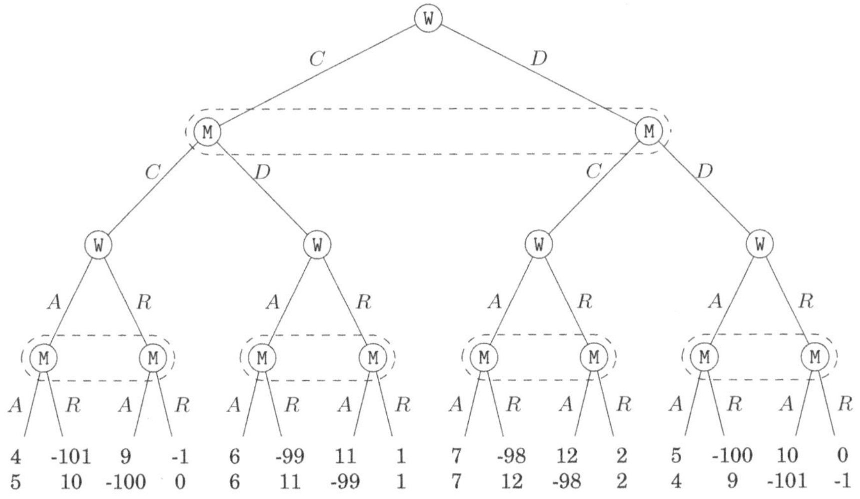

Figure 5.1: Extensive form of the game.

2. The second stage is a prisoner's dilemma and only has one Nash equilibrium (both players choose R). This is what will be played in any event by rational players in the second stage. Thus, we can remove the second stage from the computation and only focus on the first stage.

   The Nash equilibria in the first stage are to play either (C,D) or (D,C), if we consider pure strategies only. In mixed strategies, there is also another Nash equilibrium. To find it, assume C is played by the warrior with probability p and by the mage with probability q (thus, D is played with probabilities $1-p$ and $1-q$, respectively. Thus, $2(1-p) = p-(1-p)$, so $p = 3/4$. And $-q+(1-q) = 2q$, meaning that $q = 1/4$.

3. The subgame-perfect equilibria are those where one of these three Nash equilibria is played in the first stage, and then (R,R) in the second stage. These are subgame-perfect because a Nash equilibrium is played in every subgame, the only subgame other than the whole game being the second stage, where the only Nash equilibrium is (R,R).

Exercise 5.6. Chuck (C) and Dave (D) are two chipmunks searching for seeds in the wilderness, when they are attacked by a wolf. This can be represented as a two-stage symmetric game. In the first stage, they can either search an Oak (O) or a Pine (P). Their partial payoffs are represented by the following bi-matrix.

		D: O	D: P
C	O	2, 2	3, 5
	P	5, 3	1, 1

In the second stage, the wolf arrives, and they can either Flee (F) or Help (H) each other. If they both help each other, they successfully escape with just some bruises (payoff -1 for both). If only one helps, he gets devoured (payoff -100) while the other is unharmed (payoff 0). If they both flee, they manage to escape but they are both badly hurt (payoff -4).

1. Draw the extensive form of this game.
2. Find all the Nash equilibria of the two-stage game.
3. Find all the subgame-perfect Nash equilibria of the two-stage game and explain why are they subgame-perfect.

Solution

1. The normal form of the second stage of the game is:

		D: H	D: F
C	H	-1, -1	-100, 0
	F	0, -100	-4, -4

The extensive form is the tree depicted in Figure 5.2.

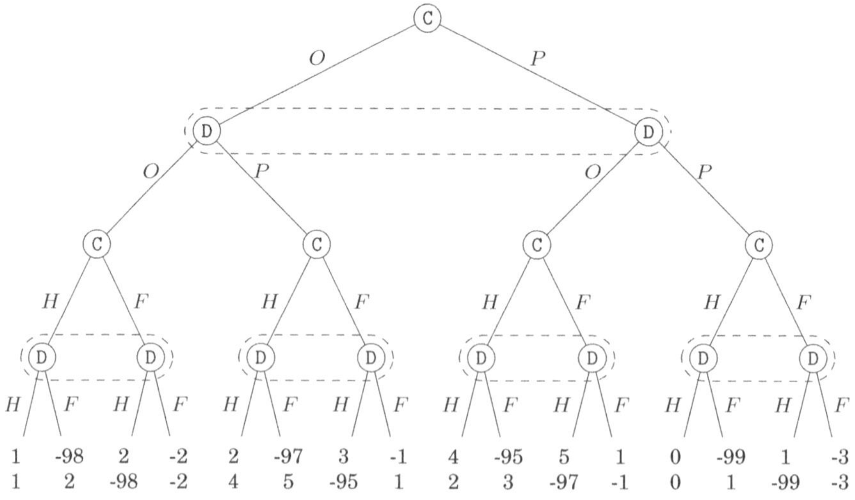

Figure 5.2: Extensive form of the game.

2. The second stage is a prisoner's dilemma and only has one Nash equilibrium (both players flee). This is what will be played in any event by rational players in the second stage. Thus, we can remove the second stage from the computation and only focus on the first stage.

   The Nash equilibria in the first stage are to play either (O,P) or (P,O), if we consider pure strategies only. In mixed strategies, there is also another Nash equilibrium where O is played with probability $p = 0.4$ and P with $1 - p = 0.6$ by both.

3. The subgame-perfect equilibria are those where one of these three Nash equilibria is played in the first stage, and then (F,F) in the second stage. These are subgame-perfect because a Nash equilibrium is played in every subgame, the only subgame other than the whole game being the second stage, where the only Nash equilibrium is (F,F).

Exercise 5.7. Two heroes, a knight (K) and a thief (T), enter the dungeon of the zombie warlock to save a princess. The king has promised a reward of 20 talents to the saviour(s) of his daughter. They reach the dungeon where the zombie warlock is waiting for them, in front of a pyre where the princess is tied to a pole. As they enter the room with their swords drawn, the zombie warlock makes an evil laugh and sets the pyre to flames; then, he turns his magic wand towards the heroes, casting a deadly spell. The two heroes must decide what to do on the spot, and cannot consult each other: attack (A) the warlock or save (S) the princess? Each of them is strong enough to kill the zombie warlock even if fighting alone. And also each of them is skilled enough to save the princess. However, if they both go for the warlock, they will easily get rid of him but the princess will die. This means no reward, i.e., utility 0 for both. If they both attempt saving the princess, it is even worse, as the warlock is left undisturbed and able to kill them, which means a utility of -100 for both. If they manage to slay the warlock and save the princess, their mission is accomplished. However, before the game ends, there is now a last stage that can be played. The hero who slayed the warlock has a further (dirty) chance to play. He can grab (G) the magic wand from the dead warlock and point it at the other hero who is busy saving the princess, killing him on the spot. This way, the reward is not split among the players, and the dead hero gets utility -100. Of course the hero who has this opportunity can also turn it down, by being a fair (F) friend to his ally: this way, the reward is split between them. K considers disgusting to kill his ally to get a bloody reward; he attributes utility -20 to this outcome. T is not so uptight about morals: getting the entire reward gives him utility 20. Splitting the reward gives utility 10 to both. Both heroes are perfectly rational and fully informed about all the details of what described before.

1. Write down the extensive form representation of this game, with the two heroes as the players, and their moves being A, S, G, F.

2. Apply backward induction to decrease one level in the game. Write down the normal form of this reduced version of the game.

3. Both players have 2 binary choices (the second one may not happen, but must be planned anyways), so they have 4 strategies as pairs of actions. E.g., "SF" means "I try saving the princess first; and if given the opportunity of betraying my ally, I will not." What kind of outcome is (AG,AG)?

Solution

1. The extensive form is the tree depicted in Figure 5.3. Since the first stage is a simultaneous-move game, it is irrelevant who is drawn first, but, without any loss of generality, consider K as the first to move.

2. If we apply backward induction, we can see that in the two intermediate sub-trees, K prefers F over G and T prefers G over F. The players can clearly

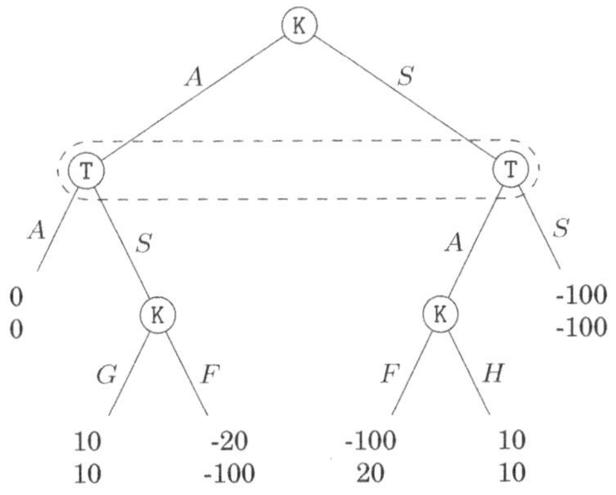

Figure 5.3: Extensive form of the game.

anticipate that. Thus, we are left with a plain static game of complete information, whose normal form is as follows.

		T	
		A	S
K	A	0, 0	10, 10
	S	-100, 20	-100, -100

For player K, playing A is strictly dominant and T is aware of that. Thus, iterating the dominance reasoning, the first stage is played as (A, S) by the heroes. Note that, in formal notation, the SPE of the game is (AF,SG), since one must specify not only that the first round is played as (A, S) but also that in the second round, if given the opportunity, K will play F and T will play G (this does not happen as it is only K who makes a second move).

3. (AG,AG) is a Nash equilibrium, albeit not subgame perfect. The result is that the princess is burned alive and the payoff is 0 for both heroes. Indeed, any situation where K threatens to play G is non-credible. However, it is an NE since no player wants to deviate: switching to play F in the second move does not change the resulting payoff of 0 for that player, while changing the first move from A to S makes the payoff -100 for that player.

Concluding remark: this exercise offers a nice interpretation from the perspective of social studies. Clearly, this does not refer to the scenario being terribly cliched and merely assigning a damsel in distress role to the only female character (very bad woke score for that). One can only appeal to the characters being based on a classic sc-fi movie, where a rogue deuteragonist is ready to shoot first. Besides, the result can be actually interpreted as a representation of how diversity of skills and morality enriches the society - in the end, even a less noble character like T can bring a contribution to the team, also because the sane part of the team is rational and aware of T's mischief. However, a team with two knights, or two thieves, would not achieve the same success of the diverse team.

Exercise 5.8. Consider a repetition for infinitely many times with discount factor δ of the stage game that is reported below in normal form. The game has a single Nash Equilibrium s $= (s_A, s_B)$, so a possible outcome of the game is that s is played forever and ever.

		A	
		M	F
B	M	1, 1	-2, 3
	F	3, -2	0, 0

1. Describe a "Tit-for-Tat" strategy that may be used by each of the player to enforce cooperation on playing M at every iteration starting from the first round.
2. For which values of δ does this cooperation take place?
3. Is this outcome a subgame-perfect equilibrium?

Solution

1. The "Tit-for-tat" is: start playing M, then play the same thing that the other player did the round before.
2. To check when this leads to cooperation, we can consider one-stage betrayal. Instead of playing M forever (that gives 1, then δ in the first two rounds), one can consider playing F, then M forever. This gives 4 in the first round, -2δ in the second. After that, the two outcomes coincide. Thus, the best response is M forever if $3 - 2\delta < 1 + \delta$, i.e., $\delta > 2/3$.
3. The "Tit-for-tat" strategy is not a subgame-perfect equilibrium because it does not give a Nash Equilibrium in all possible subgames. For example, it can lead to a death spiral if the first outcome is (M,F).

Exercise 5.9. Consider a repetition for 2 times with discount factor $\delta = 1$ of the stage game that is reported below in normal form.

A

B	M	F	H
M	7, 7	-2, 9	-1, -1
F	9, -2	2, 2	0, 0
H	-1, -1	0, 0	6, 6

1. What are the Nash equilibria of the stage game?
2. If any of these Nash equilibria are played at both stages, you obviously have a subgame-perfect equilibrium. Can you find a subgame-perfect equilibrium s $= (s_A, s_B)$ of the repeated game which is not a repetition of the Nash equilibrium of the stage game?
3. If you consider an *arbitrary* discount factor δ, what are the conditions for s to still be a subgame-perfect equilibrium?

Solution

1. Strategy M is strictly dominated by F. If we remove it, we are left with a coordination game where (F, F) has payoff 2 for both players, (H, H) has payoff 6 for both players.

   A

B	F	H
F	2, 2	0, 0
H	0, 0	6, 6

   These are the two pure strategy equilibria, and there is also a mixed strategy $1/4F + 3/4H$, with payoffs equal to $(5, 5)$.

2. For symmetry reasons, strategy s has $s_A = s_B$. Also, at the last stage s_A boils down to either F or H (or the mixed equilibrium that we call E). The normal form of the whole game is equivalent to the stage game matrix where all cells are increased of $(2, 2)$, $(5, 5)$, or $(6, 6)$. One option is to increase the top corner by $(6, 6)$ and the others by $(2, 2)$, thus creating another Nash equilibrium. This means: "play M at the first stage; then, if the outcome was (M, M) play H at the second stage; if somebody deviated play F" would work. (at the second stage you can also play E). This is a Nash equilibrium of the whole game and also of the only subgame, i.e., the last stage, where either (H, H) or (F, F) is played.
3. This technique works as long as $9 + 2\delta \leq 7 + 6\delta$ which means $\delta \geq 0.5$.

Exercise 5.10. Brother and sister play games each afternoon (for a long time ahead). They can either choose to play together in friendship (F) or to bicker (B) against each other. If they both play in friendship, they get payoff 1. If only one decides to bicker while the other plays along choosing F, they get 3 and -3 respectively. Finally, if they both end up bickering, they receive payoff 0. Future payoffs are discounted with a rate δ.

1. Represent the normal form of the stage game calling P1 and P2 the players.
2. Find a joint strategy that guarantees that the two siblings play in friendship forever, if δ is high enough.
3. Is the strategy sustainable for $\delta = 0.6$? Discuss if, for δ large enough, the outcome is subgame-perfect.

Solution

1. The set of players is {P1, P2}. Their set of strategies for the stage game is $\mathcal{S}_{P1} = \mathcal{S}_{P2} = \{F, B\}$.

		P2	
		F	B
P1	F	1, 1	-3, 3
	B	3, -3	0, 0

   The stage game is like a Prisoner's dilemma, where the only NE is (B, B).

2. This is an infinitely repeated game. There is more than one strategy that makes the two siblings play F forever. One is a Grim Trigger, where if one player deviates from (F, F), they both keep playing B forever. Another possibility is to follow a Tit-for-tat, which consists in starting playing F, and then whatever the other sibling played on the previous round.

3. The Grim Trigger can be sustained if

$$1 + \delta + \delta^2 + \cdots > 3$$

   which is for $\delta > 2/3$, thus it is sustainable for $\delta = 0.6$. On the other hand, evaluating one-stage betrayal leads to

$$1 + \delta > 3 - 3\delta$$

   so $\delta > 1/2$. Therefore, the Tit-for-tat is sustainable for $\delta = 0.6$.

   The Grim Trigger is an SPE, since in all the possible deviation from the equilibrium path result in both siblings playing the NE (B, B) of the stage game forever. The Tit-for-tat instead is just an NE that has the same equilibrium path but off equilibrium can lead to a death spiral if the outcome of a stage is (F, B) or (B, F).

6 Bayesian games

Bayesian games model strategic interactions where the involved players have some degree of uncertainty about the games that goes beyond not knowing the choice of the others, that is, they explore incomplete information.

6.1 Characteristics

In this class of games, some or all players are associated with a set of *types*. The concept of type is used to represent incomplete information that one player may have about the others. The general assumption is that players know their own types but may only estimate the types of other players. While types can be used to differentiate any aspect of the gameplay (including the way by which a dynamic game unfolds), they are most commonly employed to describe possible payoff functions for a player. Usually, players that can have a different types are often called "Bayesian players" themselves. So, a player P with a type can use different payoff functions $u_{\mathrm{A}}(\cdot), u_{\mathrm{B}}(\cdot), u_{\mathrm{C}}(\cdot), \ldots$ representing the utilities of different preference orders he puts on the possible outcomes, which is represented by its type being A, B, C, and so on. This, in turn, actually means that the other players are uncertain about what alternatives are preferred by player P, if the exact type of P is known to P only.

Generally, the assignment of types to players is represented as a decision blindly made by the virtual special player *Nature*. This means that types are random variables following a probability distribution. In principle, we should consider the joint distribution of all the types, denoted as $p(t_1, t_2, \ldots, t_{\mathrm{N}})$ for all players 1, 2, ..., N, where t_{J} is the type of player J. However, it is very frequent to assume that types are drawn independently for each player with a type, so that we can just use marginal distributions $p_1(t_1), p_2(t_2), \ldots, p_{\mathrm{N}}(t_{\mathrm{N}})$ for players 1, 2, ..., N, respectively, where

$$p(t_1, t_2, \ldots, t_{\mathrm{N}}) = p_1(t_1) \cdot p_2(t_2) \cdot \cdots \cdot p_{\mathrm{N}}(t_{\mathrm{N}}) \, .$$

While from a general perspective, in very special cases it may make sense to assume some correlation among the types, and therefore negate the independence of types, in reality independence is often assumed for tractability reasons, and also for the sake of realism, since the entire game only comprises two players (and possibly only one with a type). Thus, after drawing the types, Nature also reveals the type t_{J} to player J only.

Now, complete knowledge no longer holds as players only know the value of their own type. Clearly, this requires an adaptation of the requirements about what players do really know about the game, since we would like to express logical deductions of the players from their in-game knowledge. Therefore, in Bayesian game we need to set some common ground about what the players know, beyond what they cannot know with certainty, i.e., the types of the players with a type. A frequently adopted approach is to assume a *common prior* to all the players. This means that the joint probability distribution $p(t_1, t_2, \ldots, t_N)$ is common knowledge. Simply put, players know their own type with certainty; about others, they have an estimate (the probability distribution) of what their type could possibly be. When we can assume that this probability distribution is known by all players (the "common prior assumption"), we are able to transform games of incomplete information into games of imperfect information (in which the history of play within the game is not known to all players).

This means that Bayesian games are translated into dynamic games, and the strategies and the normal form of the game are modeled in the same ways as done there. The first move in such dynamic games is always played by Nature, and corresponds to determining the individual types. After this, we can consider that players choose an action per each of their information sets. Notably, players with a type will be aware of it and therefore choose (possibly different) actions per each one of their types, since each type leads to a separate information set. In fact, given that they have no uncertainty on their type, their information set related to the Nature's choice of their pertinence are singletons – however, they can have uncertainty on the other players' types or moves.

Very often, this leads to a *type-agent* representation of players with a type, in which their strategy is modeled like if the player suffered from split personality. In other words, these players plan in advance a strategic choice per each one of their types. This may also be combined with other multiple information sets that the players may have, e.g., due to the dynamic history of the gameplay. It may seem silly that a Bayesian player plans an action per each one of the types, since in the end the actual type determined by Nature will be only one. However, besides the general reasoning of obtaining a standard approach, this is the correct modeling choice in order to represent the *beliefs of the other players*, since they do not the type of that Bayesian player instead, and therefore can expect every sort of behavior.

Although this example is probably already old and only good for boomers, this resembles role-playing games like "One of us" or "Werewolves" where a player is dealt a secret random card/role (Nature's choice) and must play according to it. Before the game starts, that player can plan a strategy according to all possible options for his/her types. That is, the player mentally envisions something like "I will play fair if my role is that of the normal good person, but if I am one of the imposters, I will kill everyone!". Or, "If I am one of the baddies, I will act kindly so the good guys think I am on their side, and if I am good, I will launch an evil laugh to make the imposters think that I am one of them!" However, after being informed about the actual type, that player will only apply the relevant part of the strategy. Yet, the other players are in the dark about his/her type and must make all kinds of assumptions in their beliefs, e.g., is this a good character that is acting truthfully or an impostor that is pretending? Their beliefs revolve around both the *type* assigned to and the *strategy* chosen by others.

Also, an important structural property of Bayesian games is that players react to Nature's choices that are unknown to them (such as the other players' types) by taking expectations, as they would do for any other random lottery. Thus, all expectations are combined. In particular, Bayesian games give a very sensible interpretation to mixed strategies, since for a given player it is clearly identical to assume that another player with two available pure strategies A and B either (i) plays a mixed strategy where A and B are chosen with respective probabilities α and $1 - \alpha$, or (ii) is a Bayesian player with two different types t_A and t_B, and the probabilities of these types can be estimated as α and $1 - \alpha$, respectively, and finally the type-agent representation of this Bayesian player chooses A when the type is t_A and B when the type is t_B. Indeed, in both cases the response to this player choices will be planned by taking expectations over the probabilities α and $1 - \alpha$.

A common taxonomy of the strategies played by a Bayesian player, or better, its type agent representation, distinguishes between pooling or separating strategies. This classification also applies to the resulting equilibria, if they involve strategies of these kinds. The term "pooling" refers to all types of a player choosing the same strategy, so in reality the other players cannot distinguish between the types, because they all behave alike. From the perspective of an external player, the probabilities of each type of the Bayesian player can only be estimated through the prior, since no further information is gained even when the move chosen is known. Conversely, "separating" indicates that all different types play a different action, which ultimately reveal the type. Of course this requires that there at least as many available actions as types. Also intermediate cases are possible, which have different names in the literature, being called "semi-separating," "partially pooling," or simply "hybrid" – this is for example the case when a player has 3 types (say, t_{A}, t_{B}, and t_{C}) and the first two types play the same strategy A, while the last type plays another strategy C. By extension, in a pooling Nash equilibrium, all Bayesian players play the same action for all of their types, and in a separating NE they all play different actions for each of their types; once again, hybrid cases are possible, which includes all the situations where (i) some players pool their strategies and others separate, or (ii) some play a strategy that is neither pooling nor separating in their type-agent representation, or (iii) mixed strategies are also included, such as one Bayesian players play a pooling strategy and a separating one, both with non-zero probabilities.

6.2 Relevant cases

The general framework of Bayesian games can be specified to many different contexts. In particular, the solution concept used for each of these cases may change. The straightforward extension of Nash equilibrium to the Bayesian case is known as Bayesian Nash equilibrium (BNE) and it is nothing but a standard NE for a Bayesian game (which, according to the previous discussion, can be promptly translated into a dynamic game). However, similar to dynamic games, such a concept may often be insufficient to capture the rational behavior of the players in a credible way and obtain satisfactory predictions.

If the game simply implies a preliminary move of Nature for choosing the types of all involved players, and after that all choices of the players happen simultaneously, we can regard this as a *static Bayesian game*. Actually, the name is somehow misleading since it is not really a static game (there is a player, i.e., Nature, moving prior to the others). However, it is just a straightforward extension of static games of complete information, and since the type agent representation allows to remove the incompleteness of information, the BNE is more than enough to represent this case.

If the game is dynamic instead (and not just because of Nature's preliminary move), then the solution concept becomes more complex, and also the strategies available to the players may explode, since they jointly account for multiple types and different rounds of gameplay. Generally, it is tractable to manually investigate scenarios with just 2 players, only one of which is Bayesian, and only allowing two rounds of interaction.

This leads to the following relevant cases, for which the solution approach is somewhat standardized. The first is the case of a dynamic game where a first player 1 without a type moves, and this is followed by the move of a second player 2 that is instead Bayesian. This is sometimes called a *screening game* – it actually resembles the scenario in the Wizard of Oz, where an individual is hidden behind a curtain, and its actual character is unknown: is he a real wizard or just an impostor? Notably, a classic scenario where to apply this framework is the *entry game*, in which 1 also has an early termination option for the game. This somehow reduces the number of strategies and generally makes the problem more tractable - one does not have to plan the strategy of the Bayesian player 2 based on the history of play, but only on the type.

In a screening game, player 1 moves first without gaining any additional insight on player 2 and the only available information resides in the prior. Hence, 1 has only one information set (comprising all of nodes with a different Nature's choices on the type of 2, unknown to player 1) and chooses the action to play there based on the expectations, taken over the prior, of possible reactions by 2. Still, player 1 has the first mover advantage and can anticipate the outcome of the game through backward induction assuming that 2 will behave rationally. This leads to avoiding non-credible threats by player 2 and overall obtaining a subgame perfect equilibrium as the outcome. The SPE is the adequate solution concept for this kind of games, as implies that all subgames (which in this case are determined by different choices of player 1 and multiple types of player 2) are played rationally.

Conversely, one may consider a dynamic scenario analogous to a screening game, but making player 1 Bayesian instead (and player 2 is just a regular player). In this case, the SPE is found to be insufficient to describe practical solutions, as there is only one subgame (the whole game itself), because the preliminary choice of Nature being unknown to player 2 makes the information sets of this player spanning on the entire tree. Thus, all NEs are also subgame perfect, and the SPE loses its appeal. The solution is usually to resort to the Perfect Bayesian equilibrium (PBE) discussed in the next section.

One notable case where this happens is that of *signaling games*, which are especially interesting in a communication engineering context as they have close ties with information exchange, mutual information across a channel, and other aspects of information theory. The idea of a signaling game is that the Bayesian player 1 chooses an action that may or may not communicate some information about his/her type to the other player. Conversely, the information available to player 2 to react to that move may or may not be improved by observing player 1's move: in other words, player 2 might now be able to improve his/her belief about 1's type, and this should also be defined in the PBE found. In particular, it can be interesting to classify the resulting PBE as separating or pooling, since for a signaling game this can be interpreted as a characterization of a communication channel where a source sends a random variable (the Bayesian type of 1) and a destination tries to infer from the received signal. Hence, a separating PBE corresponds to a perfect information channel, where the a posteriori observation fully reveals the original value sent by the source, whereas a pooling PBE is akin to a useless channel in that the received signal gives no information, and the prior must be used instead.

A technical remark on signaling games. Very often the related literature uses the same name for two slightly different versions of such games, which, for the sake of simplicity, we will show in a case where all multiple choices of types and actions are binary. In both cases, the extensive form is usually displayed with a butterfly-like shape for better readability. The root of the tree is in the middle and implies a choice by Nature on the type of 1 which can be signal j or k, with probabilities p and $1 - p$, respectively. After that, 1's type is revealed to that player, which in turn chooses an action that can be either A or B. 2 observes just this action and plays either C or D. Payoffs are ultimately determined (there are 8 possible end nodes).

In the most general version, represented in Figure 6.1, all the actions are generic and both players have 4 strategies, but for different reasons. Player 1 has a binary choice for each type, while 2 has a no type but must make a binary choice for each of the possible moves by 1. In other words, player 2's strategy has a form XY, which is shorthand for "play X in response to player 1 playing A, and Y if 1 chooses B instead," where both X and Y are chosen as either C or D.

A slightly different version is reported in Figure 6.2, where action A is considered to be *revealing* the type of player 1. In other words, the choice of 1 is not about a specific action, but to let player 2 know his/her own type or not. The difference is subtle, and only relates to the missing dashed line between the top nodes. In this case, the top two nodes for 2 have information sets that are singletons, and backward induction immediately determines what 2 would do in such cases. As a result, there is no need to keep them into account in the strategy (even though, for the sake of correctness, they should be reported on the side). The only relevant choice of player 2 is what done in the bottom part of the tree (the non-singleton information set) where only one action is chosen. As a result, the game is smaller than before, as 1 has four available strategies, but 2 only has two.

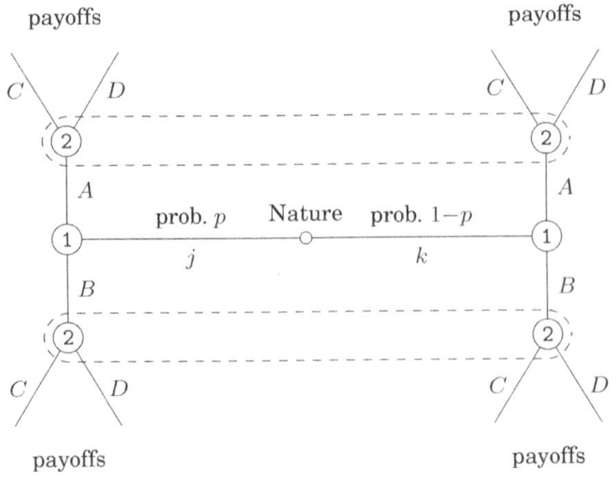

Figure 6.1: Extensive form of a standard binary signaling game.

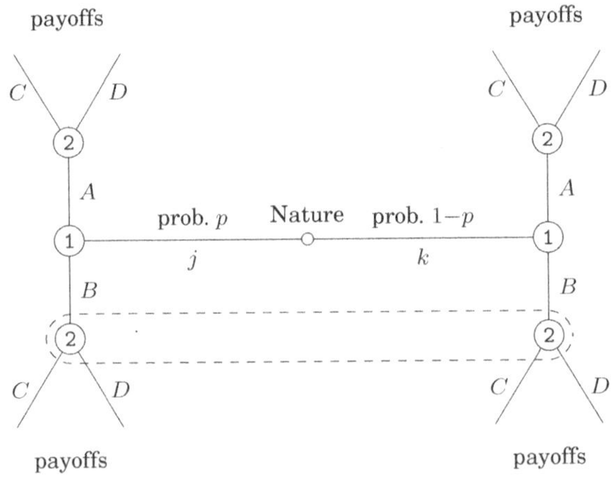

Figure 6.2: Extensive form of a binary signaling game with a revealing action.

From a modeling standpoint, one has to be careful about how the actions are defined. If the types match with the signals that player 1 can send, but player 1 is allowed to lie, we are actually in the presence of a signaling game of the first kind. If, instead, player 2 cannot be mistaken about the type of player 1 after a revealing action (but, to keep the game interesting, can instead be still in the dark after the other), we dealing with a signaling game of the second kind.

To better understand the difference, imagine that the game relates to players 1 and being a person of interest in a police case, and player 2 being the detective assigned to the case. The type of player 1 relates to having commited a crime or not. Now, a scenario of the first kind would happen if player 1 chooses a regular action, and from that action, player 2 tries to infer whether 1 is the criminal they are looking for. The detective hopes to reach a separating equilibrium, which means that 1's type is eventually inferred and 1 can be arrested or discharged. Conversely, in the second type of signaling games, the action chosen by the player is to either undergo a procedure (such as a polygraph, or a DNA swab - although they are not 100% reliable, we can assume them to be in this storytelling) that ultimately reveal its status, or refusing to take it. In that case, the revealing action A leaves no doubts about the type of player 1, while B leaves this uncertain. Of course, if 1 plays a pooling strategy on A, the type is immediately revealed and the case is solved. But also a separating strategy where the innocent type takes the swab and the guilty refuses would work (in real life, refusing to take a DNA test is often assumed for an admission of being guilty).

6.3 Perfect Bayesian equilibrium

The equilibrium concept of PBE is relevant for sequential Bayesian games, i.e., dynamic games with incomplete information, as the proper extension of the simpler concept of BNE. For the basic cases of a simultaneous-move Bayesian game, or a screening game where a Bayesian player move only at the end it is actually enough to consider the BNE or the SPE (as reinterpreted in the type-agent representation), respectively. However, the PBE is not just a generalization of the BNE in the same way as the Nash equilibrium in mixed strategies or the subgame-perfect equilibrium expand the basic idea of NE. Already the latter generalization mentioned was kinda imperfect as the definition is intrinsically different: all SPEs are NEs but not the other way around, so it is not simply an adaptation of an existing concept, but rather its expansion: think of a child class in object-oriented programming, which satisfies all the properties of the parent class plus some additional ones.

For the concept of PBE, this is pushed further in that two components are required: strategies and beliefs. To check whether the concept is understood, that is, when asked for a PBE, the answer cannot just be a strategy set, but also a set of beliefs must be specified. In a sense, this additional element was not needed for games with complete information since the idea of NE is that players have beliefs, according to which they choose a best response; and, importantly, these beliefs turn out to be correct, so that the players are actually playing a best response to the strategies of each other. In this, beliefs are already implicit in the strategy played. When incomplete information comes into play, one can no longer guarantee that the beliefs are necessarily correct and adhering to reality (this will be evident in the exercises), so they must be explicitly specified. Still, the beliefs cannot be fully detached from the strategies, as we require some degree of internal consistency, provided by sequential rationality (that must hold) and Bayes' rule to update them, from which this class of games takes its name.

In a dynamic Bayesian game, players specify their strategies so that an action is chosen per each information set, which in turn depends on the history of the gameplay. This is analogous to regular dynamic games, with the only difference that now information sets are also created based on Nature's choice of the types, which is the first move in the dynamic setup. So, per each information set, in addition to specifying a strategic choice of action, the PBE must also indicate the belief of the player concerning what node is actually the real one within that information set. Typically, this is represented through a probability distribution over the nodes belonging to that information set, which often mirrors the type of another player that is Bayesian. Think for example of the aforementioned case of a signaling game: player 2, who is not Bayesian, just observes the signal sent by 1 but this corresponds to an information set with multiple nodes. In fact, 2 may believe that 1 is of certain type and has chosen to reach that information set because 1 is of a certain type and his/her strategy dictates to play that action when being of that type, and this has multiple possible explanations (one per type of 1). In the extensive form, this corresponds to the different nodes belonging to the same information set of which only one is the actual location of the game history: while 1 knows this, 2 does not and is undecided.

Still, one can easily see the analogy between this situation and decision theory in classic scientific reasoning: player 2 is not entirely in the dark, but can apply reasoning skills to infer some conclusions and quantify what he believes to be the posterior probability of each node, after observing the action chosen by 1. Specifically, the players must create a *system of beliefs* that satisfies the aforementioned condition of sequential rationality (each action must locally maximize the expected payoff for the undertaking player, given the beliefs), and beliefs are updated following Bayes' rule.

In practice, beliefs within an information set h spanning over n nodes (that differ from one another in representing the unknown type of a Bayesian player) are an n-tuple of non-negative values $\mu_1, \mu_2, \ldots, \mu_n = 1 - \mu_1 - \mu_2 - \cdots - \mu_{-1}$, thus summing to 1 so as to represent a probability distribution. When updating them, it is generally distinguished between nodes on and off the equilibrium path. In the probabilistic context of Bayesian games, a node is on the equilibrium path if it is reached with positive (i.e., non-zero) probability. For these nodes, Bayes' rule in the case mentioned above implies that the belief μ_j for the jth node (which denotes the jth possible type of a Bayesian player) is the ratio between the marginal probability that the information set is reached when the type of the Bayesian player is j and the unconditional probability that the information set is reached at all. The latter is usually computed through the total probability theorem as the sum of all possibilities. In other words:

$$\begin{aligned} \mu_j &= \frac{\Pr[h \text{ is reached } \& \text{ the Bayesian type is } j]}{\Pr[h \text{ is reached}]} \\ &= \frac{\Pr[h \text{ is reached } \& \text{ the Bayesian type is } j]}{\sum_{k=1}^{n} \Pr[h \text{ is reached } \& \text{ the Bayesian type is } k]} . \end{aligned} \tag{6.1}$$

However, Bayes' rule as expressed by (6.2) only works if $\Pr[h \text{ is reached}] > 0$, otherwise we are dividing by 0. When $\Pr[h \text{ is reached}] = 0$ we consider to be off the equilibrium path, and this result in different requirements proposed in the literature as regarding what to do in this case. The usual approach is that of a *weak* PBE that implies that the values of the μ_j's in this case are arbitrary. Some authors impose stronger conditions, to obtain a more selective definition of PBE, but in many practical applications this can be often seen as nitpicking. At any rate, consider that the values of the μ_j's being arbitrary does not mean that they can be any number, in fact they must still represent probabilities and therefore they cannot be negative or larger than 1 and they must also sum to 1, meaning that they cannot be all 0.

Exercises

Exercise 6.1. Little Stevie really hopes to receive a space train (T) for Christmas. His parents have already bought him a sweater (S) but they tell him that he might receive T instead if he behaves as a good boy (G) and not a naughty one (N). Stevie's behavior ultimately depends on whether Santa Claus exists or not, which he evaluates as having probability p. If Stevie is naughty, he will not receive any gift, regardless of Santa Claus existing or not. Stevie's and the parents' utility will be 0 and -10 respectively. If Stevie is good, he will either receive S or T, according to the decision made by his parents. If Santa exists, Stevie can receive the space train at no cost for his parents. If Santa does not exist, Stevie can receive T only if the parents pay for it. Stevie's utility for S and T is -10 and $+40$, respectively. The parents' reward when giving S and T to their son is 0 and $+20$. However, if they have to buy the space train themselves, subtract 50 from their utility. Also note that, being adults, the parents know whether Santa exists or not; on the other hand, they also know the value of p estimated by their son (that is, the prior is common knowledge).

1. Represent this Bayesian game with player 1 being Stevie and player 2 being his parents in extensive form.
2. Represent this Bayesian game in normal form, with a type-player representation of player 2.
3. Find the Bayesian Nash equilibria of the game in pure strategies if $p = 0.9$. Discuss what kind of equilibrium concept is appropriate for this game, and elaborate whether the pure BNEs found satisfy it.

Solution

1. The extensive form is the tree depicted in Figure 6.3.

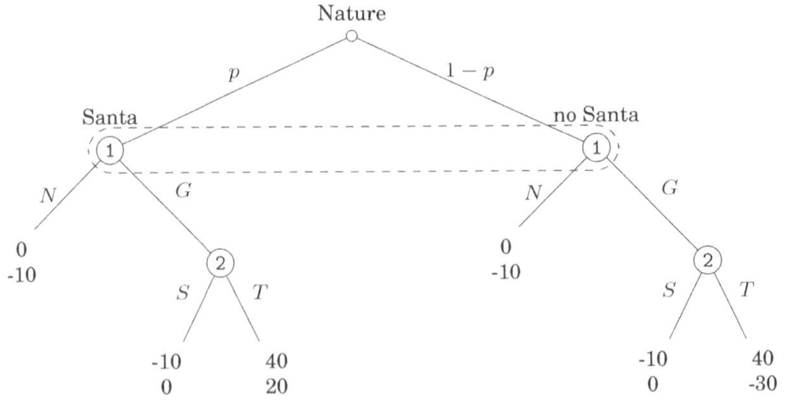

Figure 6.3: Extensive form of the Bayesian game.

Since Stevie has no type, this is a Bayesian game known as a *screening* game (i.e., a dynamic game where the player moving first has no type). Players are Stevie as 1 and his parents as 2. The root node is Nature's move about Santa Claus's existence with p and $1-p$. Below it, Stevie moves without distinguishing between the two nodes.

2. Type-agent representation of player 2 decides a strategy as a pair of actions (Santa – no Santa). For example, ST means they give a sweater if Santa exists, they give the space train if Santa does not. This leads to the following bi-matrix.

1 \ 2		SS	ST	TS	TT
1	G	-10, 0	40-50p, -30+30p	50p-10, 20p	40, 50p-30
	N	0, -10	0, -10	0, -10	0, -10

3. If $p = 0.9$ the matrix above becomes

1 \ 2		SS	ST	TS	TT
1	G	-10, 0	-5, -3	35, 18	40, 15
	N	0, -10	0, -10	0, -10	0, -10

Since this is a screening game, the proper equilibrium concept to consider here is that of subgame-perfect equilibrium. The Bayesian Nash equilibria in pure strategies are (N,SS), (N,ST) and (G,TS). However, only the last one is an SPE, which is clear from it being the result of backward induction. The others contain a non-credible behavior by the parents (Santa Claus exists, yet they give the sweater when the train comes for free).

Side note: this game emphasizes some aspects for what concerns Bayesian games and also psychological implications on human behavior. First of all, there will be spoilers ahead. So keep reading at your own risk, or turn the page, now. The spoiler is: Santa Claus does not actually exist. Sorry if nobody told you until now. In light of this, consider that the game is Bayesian but the player with a type (player 2, i.e., the parents) has no essential difference. The type of that player relates to a different individual, that still influences the player to which we attribute said type. Otherwise said, a type does not necessarily describe a characteristic of the typed player, but rather an aspect of the world that has consequences on the behavior of said player. In particular, here the type relates to Santa Claus being real or not – which indeed affects the behavior of the parents, since if Santa Claus was real, Christmas' gifts were a lot cheaper.

Also, the relevant aspect of a type is all in the beliefs (and what better belief than the one of a kid about Santa). It does not even matter whether they are correct or not, in this problem all that counts is that Stevie is convinced that Santa Claus exists, at least most likely – he has some doubts but he puts the overall probability of Santa Claus being real to a relatively high value of $p = 90\%$. Also the belief of the parents about Santa does not matter either, what is important is that they are aware that Stevie has this estimate for p, in other words, p must be a common prior (so also Stevie knows that his parents know and so on). In studying this Bayesian game, a suspension of disbelief is required in that the analysis follows from Stevie's beliefs and not from reality.

Finally, it is rather evident that the SPE of the game above is (G,TS) because Stevie believes that Santa Claus is most likely real, and this leads him to behave as a good kid. You can easily verify that a lower value of p will change the SPE: the parents will behave in the same way (after all, their moves are certain, once their type, which is in reality the real nature of Santa, is known), but Stevie will be naughty. You can draw your own conclusion about why the myth of Santa Claus was invented, then.

Exercise 6.2. At the saloon, a young cowboy Y is insulted by a black hat outlaw O, that is feared to be the fastest gun in the state. Y can let it slide (S): in this case, Y gets utility -20 and O gets utility 0. Or, Y can challenge O to a duel (C), in which case O can either apologize (A) or accept the duel at high noon (D). If O apologizes, Y gets utility 10 and O gets utility -10. The outcome of the duel depends on whether O is really a sharpshooter or not. If O is a sharpshooter, Y has no chance of winning the duel. The payoffs are -100 for Y and 20 for O. If O is just pretending to be a fast gun, then the duel is uncertain: the probability of winning is 0.5 for both Y and O, and whoever wins gets utility 20 and whoever loses gets -100. The probability of O being a sharpshooter is a common prior equal to p.

1. Represent this Bayesian game with players Y and O (in this order) in extensive form.
2. Represent this Bayesian game in normal form, with a type-agent representation of player O.
3. Find the Bayesian Nash equilibria of the game in pure strategies if $p = 0.2$. Discuss what kind of equilibrium concept is appropriate for this game, and elaborate whether the pure BNEs found satisfy it.

Solution

1. This is a screening game, since the player moving first has no type. It is basically an entry game: instead of a company risking its entrance in a market dominated by an incumbent, we have a young cowboy considering whether to challenge the outlaw to a duel.

   Note that there are also some lotteries involved, to determine the outcome of the duel (one of which, when O is a sharpshooter, is degenerate). However, from the previous chapters we already know that it is sufficient to take the *expected payoffs* in this case. Since Bayesian games always take expectations on Nature's choices, these expectation will stack up with those of the lotteries, but due to linearity of expectations, we can simply take a single big expectation when computing the payoffs.

   The extensive form is the tree depicted in Figure 6.4. The root node is Nature's move about O being a sharpshooter with p and $1-p$. Below it, Y moves without distinguishing between the two nodes.

2. Type-agent representation of player O decides a strategy as a pair of actions (sharpshooter – liar). For example, DA means O does not duel if he is a sharpshooter, while it does if he is not. This leads to the following bi-matrix.

O

		AA	AD	DA	DD
Y	C	10, -10	50p-40, 30p-40	-110p+10, 30p-10	-60p-40, 60p-40
	S	-20, 0	-20, 0	-20, 0	-20, 0

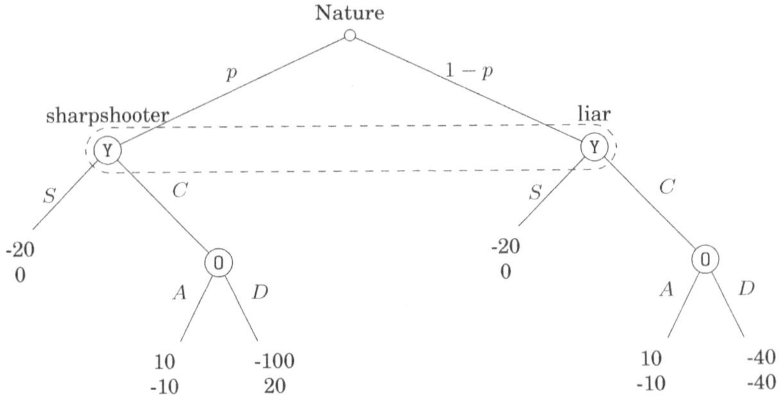

Figure 6.4: Extensive form of the Bayesian game.

3. If $p = 0.2$ the matrix above becomes

		O			
		AA	AD	DA	DD
Y	C	10, -10	-30, -34	-12, -4	-52, -28
	S	-20, 0	-20, 0	-20, 0	-20, 0

The Bayesian Nash equilibria in pure strategies are (S,AD), (C,DA) and (S,DD). Since this is a screening game, it is enough to consider the concept of subgame-perfect equilibrium. It turns out that only the first one is an SPE, as visible from it being the result of backward induction. The other two contain a non-credible threat from O, that is, to accept the challenge of a duel if he is not a sharpshooter, which is non-believable (also, the second one has the further irrational action where O apologizes when he is a sharpshooter, which is against his utility). In the right hand subtree, O will always choose A over D and Y can anticipate it.

So the only SPE is (C, DA). In this case, we do not need to consider the concept of PBE, but if we do, this can only confirm this result. Indeed, since Y moves first, he must use the prior to discriminate among the nodes, so his beliefs are just p and $1 - p$ anyways. In case of the equilibrium (C,DA), Y plays rationally and the equilibrium path involves the rational choices of player O (to apologize if weak, to accept the duel if sharpshooter). The beliefs of O are always 100% in the nodes where he moves, since he knows his own type. Instead, in the other two BNEs, the choices of O are off the equilibrium path. Therefore, his beliefs can be arbitrary, but they cannot be zero, hence showing that in this evolution of the game, O has planned an irrational strategy and these two BNEs do not extend to a PBE.

Exercise 6.3. A broadcasting system S has bought a license from the government to transmit on some radio channels, but it is not using all of them all the time. An opportunistic transmitter T wants to operate on a channel akin to those licensed to S. Thus, T can choose between: (B) buy a license from the government for another channel not licensed to S; (A) access one of the channels licensed to S, hoping that S does not mind. If T plays B, it gets utility -10 for the channel cost, and S gets utility 100 because it is left alone undisturbed. If T plays A, S will see some additional interference, and it can either (C) combat it by raising its transmission power, or (D) do nothing. If S plays D, it gets utility 80 because of the disturbance; T is able to transmit and gets utility 20. If S plays C, T always gets utility -20; the utility of S depends instead on whether high power supply are available to S or not. If high power supply are available, then S gets utility 90. If not, S gets utility 50. The opportunistic transmitter T does not know whether S has high power supply or not, but estimates this to have probability p; this value is common knowledge.

1. Represent this Bayesian game with players T and S (in this order) in extensive form.
2. Represent this Bayesian game in normal form, with a type-player representation of player S.
3. Find the Bayesian Nash equilibria of the game in pure strategies if $p = 0.7$ and discuss what kind of equilibrium concept is appropriate for this game.

Solution

1. This game is a Bayesian entry game. Also, it is a quite common model for opportunistic transmission in *cognitive* networks, which is a very popular research area (both in itself and for its game theoretic approaches). If you are interested in studying it, you do not need much more beyond what discussed in this exercise, as most of the analyzed interactions precisely involve Bayesian situations of competing transmitters, even though the actual development of the game may be more complicated or have more players.

   The extensive form of this game is the tree depicted in Figure 6.5.

2. Type-player representation of player S decides a strategy as a pair of actions (high power – low power). For example, DC means S does not combat interference if it has high power, while it does if power is low. This leads to the following bi-matrix.

S

T		CC	CD	DC	DD
	A	-20, 50+40p	20-40p, 80+10p	40p-20, 50+30p	20, 80
	B	-10, 100	-10, 100	-10, 100	-10, 100

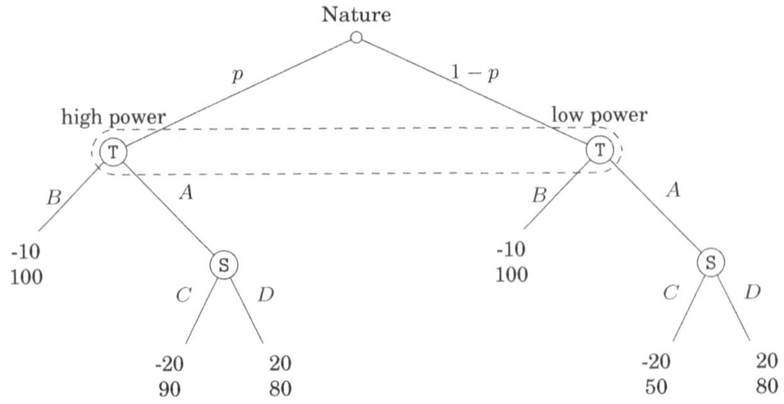

Figure 6.5: Extensive form of the Bayesian game.

3. If $p = 0.7$ the matrix above becomes

S

		CC	CD	DC	DD
T	A	-20, 78	-8, 87	8, 71	20, 80
	B	-10, 100	-10, 100	-10, 100	-10, 100

The (Bayesian) Nash equilibria are (A,CD) and (B,CC). Since this is a screening game, the appropriate solution concept is that of subgame perfect equilibrium. Of these BNEs, only (A,CD) is an SPE. The other contains a non-credible behavior by the broadcasting system, that raises its power even when it does not have a high power supply.

Exercise 6.4. Franco (F) and Gino (G) are henchmen who have been paired together by their padrino to send some warning to a rival gang. They can, independently and without preliminary agreement, act in a Conspicuous (C) or Inconspicuous (I) way. If they both are inconspicuous, the warning is ineffective and the padrino will not give any reward to the henchmen. If at least one is conspicuous, the rival gang gets a scary warning and the padrino rewards the henchmen; this reward has value 10. However, if they both are conspicuous, the warning is too loud and the police arrests them. As a result, their reward is -30. Gino is about to prepare for this mission when he hears a rumor: Franco might be an undercover agent of the police. If this is true, Franco's preferences are different. He *wants* them to be caught, so if they both are conspicuous, the reward of the undercover Franco is 20. Otherwise, the undercover Franco always gets utility 0. Gino estimates the likelihood that Franco is police with probability p. Franco clearly knows whether he is an undercover officer or a real henchman, but he is also aware of Gino's suspicion – and the value of p is common knowledge.

1. Write down the extensive form representation of this game.
2. Write down the normal form of a type-agent representation of this Bayesian game (where all typed players have strategies represented as n-tuples of actions, one per type).
3. Find out the Bayesian Nash equilibria of this game for $p = 0.3$ and discuss whether they are subgame-perfect or what are the corresponding PBEs.

Solution

1. The extensive form is a tree. The first node is Nature, deciding whether F is undercover or not, with probability p and $1 - p$ respectively. Below that, the game can be represented in two ways. One way, shown in Figure 6.6, consists in F moving first. He does distinguish between the two nodes below, since he knows whether he is undercover or not. Afterwards, G moves without knowing if F is police, nor what action is chosen by F. Therefore, all the four nodes of G belong to the same information set.

   Another possibility is to have instead G moving first without knowing whether it is the left-hand or right-hand subtree, and afterwards F knowing Nature's choice, but not G's move. This "swap" is allowed whenever the game is played simultaneously.

2. Player F has a type, so his type-agent representation has four strategies (two types, two actions): CC, CI, IC, II. The normal form representation is

F

		CC	CI	IC	II
G	C	-30, 50p-30	10-40p, 10p+10	40p-30, 30p-30	10, 10-10p
	I	10, 10-10p	10p, 0	10-10p, 10-10p	0, 0

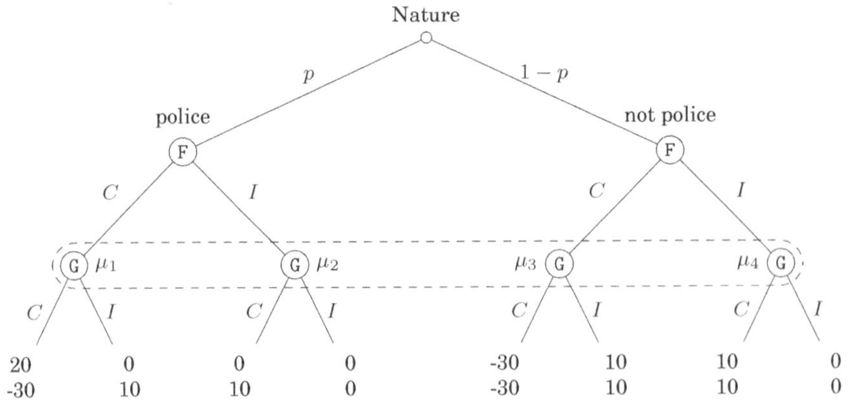

Figure 6.6: Extensive form of the Bayesian game.

3. If $p = 0.3$ this results in

		F			
		CC	CI	IC	II
G	C	-30, -15	-2, 13	-18, -21	10, 7
	I	10, 7	3, 0	7, 7	0, 0

The best responses of G to F's strategies are: I, I, I, C, respectively. The best responses of F to G's strategies are: CI (to C); both CC and IC (to I).

The Bayesian Nash equilibria in pure strategies are thus (CC, I) and (IC, I). Note that II is weakly dominated for F. Since this is the only choice of F where the best response for G is to play C, G always plays I. Thus, we can say that, because of the suspicion about F, G always plays inconspicuously. But F can anticipate that and, if he is a real henchman, he plays C, while the undercover F is indifferent between C and I, which give him utility 0 anyways.

Thus, since the "police" type of F is indifferent between C and I, the BNEs of this game, also including mixed strategies, are a family (meant as a set, not as a mafia clan, in case you are wondering) of infinitely many equilibria, which includes all the cases where:

- G chooses I
- the "not police" type of F always plays C
- the "police" type of F plays a linear combination $\alpha C + (1 - \alpha)I$, with $\alpha \in [0, 1]$.

Since this game is a "static Bayesian," this is already a good description of the resulting equilibrium. However, when considering the extensive form, all BNEs in the family described above are subgame perfect, since there is a single subgame. They can also be extended to *perfect* Bayesian equilibria if we consider the system of beliefs and explicitly describe it. For F, all information sets are singletons. For G, there is a unique information set, with four beliefs $\mu_1, \ldots, \mu_4$ as specified in Figure 6.6.

Since the "not police" of F type never plays I, $\mu_4 = 0$, and therefore $\mu_3 = 1 - p = 0.7$, based on the prior. Instead, the first two nodes are reached with probability α and $1 - \alpha$ by the "police" type of F, which is conditioned to happen with probability p. Thus, $\mu_1 = 0.3\alpha$ and $\mu_2 = 0.3(1 - \alpha)$.

Exercise 6.5. John (J) sees his friend Camilla (C) downtown. Actually, John is near-sighted and he is not really sure whether that girl is Camilla or not. The probability that C is actually Camilla is estimated as p, and this is a common prior. John can either ignore (I) the girl or greet (G) her. If J chooses I, the game end with an ashamed J getting payoff -20; in this case, C gets -50 if she is really Camilla, otherwise she gets 0. If J chooses G, then the girl has two choices, wave back (W) at John, or stay silent (S) and go ahead. If the girl waves back: J always gets payoff 10 (regardless of that being Camilla or not); the girl gets 30 if she is the real Camilla, otherwise she also gets 10. If the girl stays silent: if the girl is Camilla, J gets -50 and C gets 10; otherwise they both get 0.

1. Represent this Bayesian game with players J and C (in this order) in extensive form.
2. Represent this Bayesian game in normal form, with a type-agent representation of player C.
3. Find the Bayesian Nash equilibria of the game in pure strategies if $p = 0.3$ and discuss what kind of equilibrium concept is appropriate for this game.

Solution

1. This is a Bayesian entry game. The extensive form is the tree of Figure 6.7.

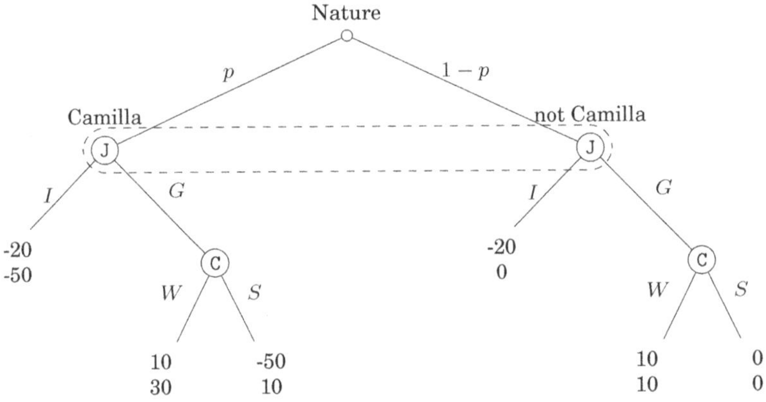

Figure 6.7: Extensive form of the Bayesian game.

2. Type-agent representation of player C decides a strategy as a pair of actions (Camilla – not Camilla). For example, SW means C does not wave if she is Camilla, while she does if she is not. This leads to the following bi-matrix.

C

J		SS	SW	WS	WW
	I	-20, -50p	-20, -50p	-20, -50p	-20, -50p
	G	-50p, 10p	10-60p, 10	10p, 30p	10, 20+10p

3. If $p = 0.3$ the matrix above becomes

C

J		SS	SW	WS	WW
	I	-20, -15	-20, -15	-20, -15	-20, -15
	G	-15, 3	-8, 10	3, 9	10, 23

The only Bayesian Nash equilibrium is (G, WW). This is also subgame-perfect, since W is best response to G, regardless of whether the girl is Camilla or not. Since this is a Bayesian entry game of the screening kind, discussing subgame-perfect equilibria is sufficient. If one wants to translate this into a PBE, it is sufficient to explicitly include the beliefs, but they are just the prior for the only non-typed player J, that moves first and therefore has no chance to observe C's move to update them.

Exercise 6.6. Mark (M) needs to buy a new pair of trousers at the shop owned by Nadia (N) within the week. The price at the beginning of the week is €50. Mark can buy the trousers immediately (I) or wait until the end of the week (W) hoping that Nadia will open a promotion sale. Indeed, Nadia has the option of keeping the price identical (K) for the whole week or lowering the price (L) to €30 at the end of the week, and she is pondering what to do. Nadia's choice depends on the original cost of the trousers that she paid, which she knows (but Mark does not). With probability $p = 0.75$, the trousers cost €20, so she still makes a profit out of the lower price. With probability $1-p$, she paid €35 for the trousers, so she is not happy about a discount. The value of p is a common prior. Mark and Nadia's respective utilities are their net gain in €, considering that Mark's evaluation of a new pair of trousers is €100, and having the transaction made at the end of the week implies lower utilities by a factor $\delta = 0.8$ for both of them (note: δ is a discount factor, but applies to the utilities, it has nothing to do with the discount applied on the sale price). Finally, if Nadia is able to afford doing a discount without an economic loss, she also gains an equivalent of €20 due to increased customer satisfaction (Mark will return to her shop to buy more clothes).

1. Model this as a Bayesian game in extensive form, clarifying your choices as for: which player(s) have a type, who moves first, what are the end payoffs.
2. Give the normal form of this game with a type-agent representation of the typed players, i.e., their strategies are n-tuples of actions (one per type).
3. Find all Bayesian Nash equilibria in pure strategies and discuss what kind of equilibrium concept is appropriate for this game.

Solution

1. The extensive form of this Bayesian entry game is shown in Figure 6.8.

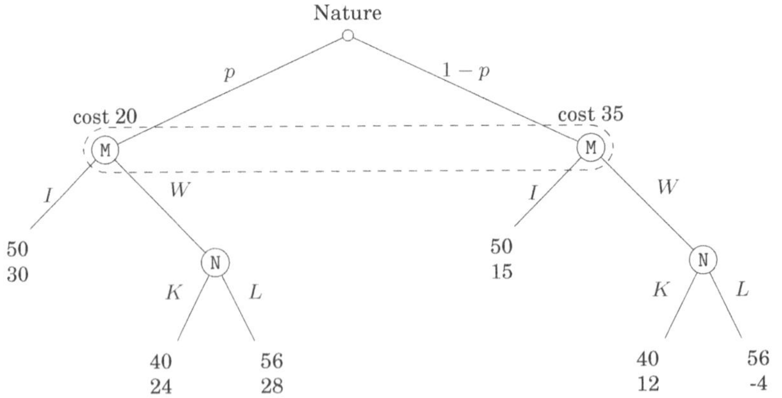

Figure 6.8: Extensive form of the Bayesian game.

Nadia has a type, i.e., the original cost of the trousers, known to her, but not to Mark. Thus, the first move is Nature's deciding about this cost with probability p or $1 - p$. Then Mark moves without distinguishing between the two nodes that belong to the same information set: if he decides to play I, the game ends without Nadia having a say in this. If Mark plays W, then Nadia is allowed to move, and she is able to tell the difference of the two nodes and her available moves are K and L in both cases.

2. A type-agent representation of player N implies that her strategy is a pair of actions: what to do when the cost is €20, and what to do when the cost is €35. Both elements can be either K or L. For example, KL means Nadia chooses K if the cost is €20 and L if the cost is €35. This results in the following bi-matrix a generic value of p.

		N			
		KK	KL	LK	LL
M	I	50, 15+15p	50, 15+15p	50, 15+15p	50, 15+15p
	W	40, 12p+12	56-16p, 28p-4	40+16p, 16p+12	56, 32p-4

For $p = 0.75$, the payoffs become:

		N			
		KK	KL	LK	LL
M	I	50, 26.25	50, 26.25	50, 26.25	50, 26.25
	W	40, 21	44, 17	52, 24	56, 20

3. This Bayesian game has three (Bayesian) Nash equilibria in pure strategies: (I, KK), (I, KL), (W, LK). Since this is a screening game, subgame-perfect equilibria is the appropriate equilibrium concept to discuss. Only the third BNE is also an SPE, as the first two involve two non-credible moves by Nadia (playing K when the price is €20). This irrational choice has no effect on the Nash equilibrium, as it is never played (Mark actually chooses I), but there are subgames off-equilibrium in which this implies that a Nash equilibrium is not obtained.

Exercise 6.7. Janine (J) is going for a working trip to a small neighborhood. She knows there is located a nice sandwich stand (S), that is the only option for having lunch there. J can consider whether to bring food from home (H) or eat a sandwich at the stand (E). However, the stand has two workers, Alfred (A) and Brianna (B), and J does not know who is there today; she just knows that only either of them is working (they never work together) and A and B there 80% and 20% of the time, respectively. This can be seen as an independently drawn random variable, and this fact is common knowledge for all the players. Moreover, once J reaches the end of the trip, if she tries buying a sandwich, the worker (whoever they are) can actually give her one (G) or deny it to her with an excuse (D). Actually, action D is very satisfying with worker B, that for some reasons hates J. Thus, a game can be played between two real players, J and S, with a further virtual player "nature" deciding who is in service at the stand. This decision is made known to S but not to J. Then J decides before leaving for the trip whether to bring sandwiches or not. Then, it is S turn to play after observing J's move. If J plays H, the game actually ends and the payoffs of J and S will be $+2$ and 0, respectively. If J plays E, the payoffs are determined by the action of player S. If S gives J a sandwich (action G), J's payoff will be 5, otherwise (action D) J gets 0. The reward of player S when selling a sandwich is 1, and of course 0 if the sandwich is not sold to J with an excuse. However, the payoff of player S is increased to 5 in the specific case that the worker at the stand is B and the sandwich is *not* sold.

1. Represent this problem as a game in extensive form.
2. Write down the normal form of the game with a "meta-player" S in agent-type representation.
3. Find the Bayesian Nash equilibria in pure strategies, and discuss whether they are subgame-perfect.

Solution

1. The extensive form of the game is the tree depicted in Figure 6.9. The root is Nature, deciding whether Alfred or Brianna are at the stand. Then J moves between H and E. Whenever J chooses H, regardless of who is at the sandwich stand, payoffs are (2, 0) and the game ends there. If J chooses E, payoffs are (5, 1) for G and (0, 0) for D if A is at the stand; if B is at the stand, the latter pair is modified to (0, 5).
2. Type-agent representation of player S decides a strategy as a pair of actions: what Alfred does and what Brianna does. Both elements can be either G or D. For example, GD means Alfred gives Janine a sandwich, but Brianna does not.

S

J		GG	GD	DG	DD
	H	2, 0	2, 0	2, 0	2, 0
	E	5, 1	4, 1.8	1, 0.2	0, 2

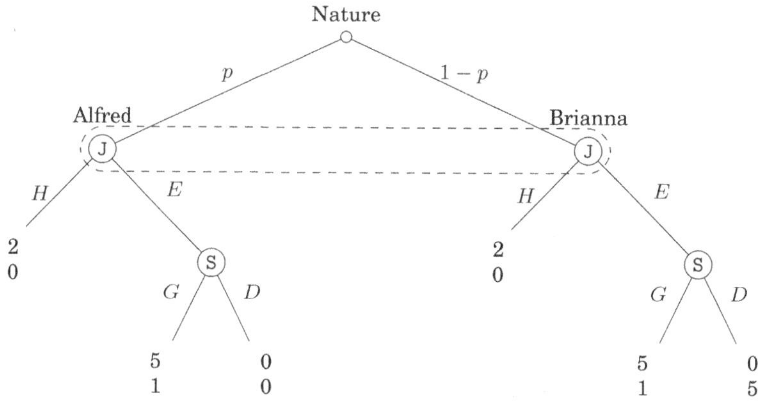

Figure 6.9: Extensive form of the Bayesian game.

3. There are three BNEs: (E, GD), (H, DG) and (H, DD). Only (E, GD) is subgame-perfect, being the result of backward induction. The other two have a non credible behavior of player S. Notably, the SPE can be extended to a PBE by also including beliefs for J, which in this case are just the prior (the game is a screening game).

Exercise 6.8. Italian tourist Riccardo (R) desires to visit the country of Bravonia. Due to an ongoing pandemic, Bravonia allows people in only if they undertake a PCR test with a very annoying nasal swab, to show that they are not infected. R took a blood test in Italy very recently so he knows whether he is infected or not, but the test result is in Italian, a language that the Bravonese custom officer (B) does not understand. When reaching the Bravonese border, R decides whether to accept the swab (S) or try talking his way (T) to convince B that he is not infected. R can choose T even if he is actually infected, since R can lie of course. If R chooses S, then B becomes aware of whether R is infected or not. After R's choice, B can admit him in the country (A) or deny him access (D). The utility of R is 20 or 0 when allowed to enter the country or not, respectively. But these values are decreased by 10 if R undergoes the annoying nasal swab. The utility of B is 0 when deciding not to admit R. If R is admitted, B's utility is -50 or 5, depending on whether R is infected or not, respectively. Finally, B can only estimate that the average Italian tourist is infected with probability $p = 0.2$, which is a common prior. All of the previous information is common knowledge among the players, except for R being infected or not, which is known only to R himself.

1. Write the game in extensive form.
2. Write the game in normal form with a type agent representation of R.
3. Find the Perfect Bayesian Equilibria of this game and discuss whether they are pooling or separating or hybrid.

Solution

1. The extensive form is the tree depicted in Figure 6.10.

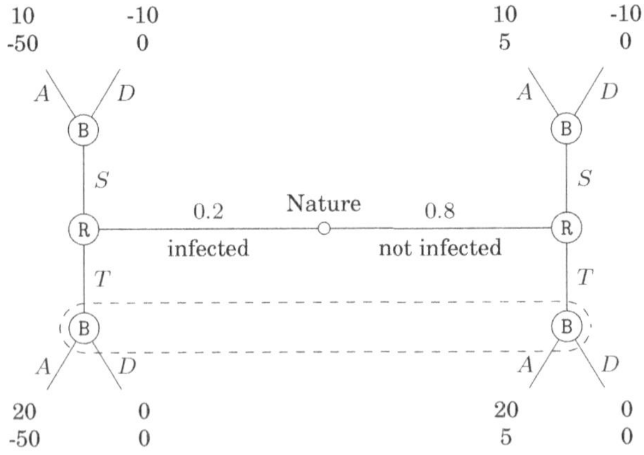

Figure 6.10: Extensive form of the Bayesian game.

The root of the tree is Nature, deciding whether R is infected (probability $p = 0.2$) or not (probability $1 - p = 0.8$). After this, R moves to nodes where it is B's turn, but if R chooses S, the resulting nodes belong to separate information nodes, whereas if R chooses to play T, they belong to the same information set.

Since S is a revealing action, the upper part can be simplified thanks to sequential rationality, since B will always play D and A, respectively, in his upper nodes. The resulting tree is shown in Figure 6.11.

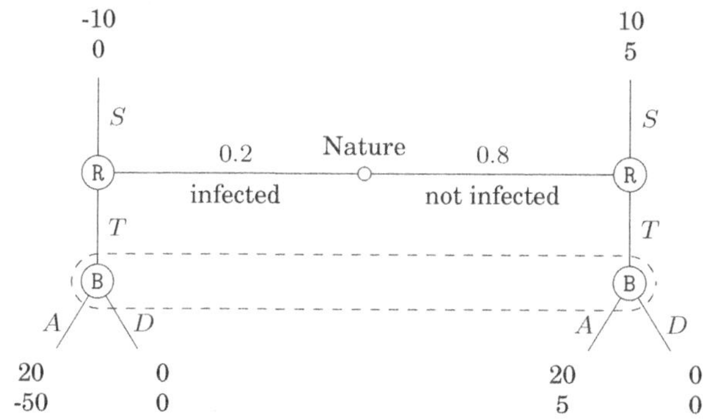

Figure 6.11: Simplified extensive form of the game after applying sequential rationality to the upper nodes.

2. The normal form must consider four possible strategies of R, being pairs of binary S/T values, representing "what to do if infected/what to do if not infected". For example, ST means that R takes the swab if infected and does not take if not infected. For player B, there is only one action to consider, i.e., the one taken in the information set after R plays T. This can be either A or D. The result is the following bi-matrix.

B \ R	SS	ST	TS	TT
A	4, 6	4, 14	-6, 12	-6, 20
D	4, 6	0, -2	4, 8	0, 0

3. The only Nash equilibrium of the game is (TS, D). To check if this is a PBE (clearly it is a *separating* PBE since the Bayesian player reveals his type: choosing T implicitly implies that he is infected) we must also set: (i) the belief of B when seeing T, which is that R is infected with probability $\mu = 100\%$ and (ii) the action chosen when the infected type chooses to reveal. The latter happens off the equilibrium path, since the infected type never reveals and strategies SS and ST are strictly dominated. Still, if the infected type chooses S, then B clearly plays D.

Exercise 6.9. John (J) and Mary (M) just got engaged and they are considering to buy each other gifts for Christmas and Valentine's Day. Choosing to buy a gift (G) costs 5 in terms of money and effort, but gives a pleasant benefit of 12 to the other. Of course, if one does not buy a gift (N), this leaves the payoffs of both players unchanged. This "Christmas" – "Valentine" situation is modeled as a repeated game of *two* interactions without discount ($\delta = 1$). In each of them, J and M choose their (G/N) action simultaneously and unbeknownst to each other, but with memory of the past. M is known for being a *strategic* player that just acts selfishly. With probability p, J is a "Tit for Tat" player that always starts with G in the first round but then plays in the second round what M did in the first round. With probability $1 - p$, J is also a strategic player. The value of p is common knowledge.

1. Compute the total payoffs of both players if $p = 0$.
2. What is the probability that J plays N in round 1?
3. What condition on p you can impose for having M playing G in the first round?

Solution

Preliminary note: despite involving Bayesian types, this is basically just a prisoner's dilemma repeated twice. The collaborative action is G, while N represents a defection. As discussed below, the role of the Bayesian type is to introduce uncertainty of the behavior of the players, so that some degree of collaboration can be established. But overall, the SPE is sufficient as a solution concept for this game.

1. If both players were declared as strategic, as in the standard repeated prisoner's dilemma, we know that collaboration is not possible because this is a finitely repeated game. This is exactly the case when $p = 0$ and thus both players defect (i.e., play N) both times. Their payoffs are 0. This happens since the stage game only has one NE, that is, (N, N), game $\mathbb{G}(2, \delta = 1)$ only has 1 SPE, which is the repetition of the stage-game NE.

   To formally represent the SPE, we must remind that each strategic player must specify 5 actions (the first action, and then the response to each of the four outcomes), which gives the involuted notation of ($NNNNN$, $NNNNN$).

2. As discussed above, a strategic J plays N, while a Tit-for-Tat J plays G in the first round. As a consequence, J plays N in the first round if and only if he is strategic, i.e., with probability $1 - p$.

3. This part is actually the really interesting one of the exercise. As the question implies, there is an option that player M plays G (i.e. collaborates) in the first round despite being strategic. Note that, in the last round, she will surely play N. But if she plays N in the first round, too, the two

outcomes are (G, N) in round 1 and (N, N) in round 2 if J is tit-for-tat, or (N, N) for both rounds if J is strategic.

Instead, if she collaborates with J in the first round by playing G, the possible outcomes are:

- (N, G) and (N, N) if J is strategic;
- (G, G) and (G, N) if J is tit-for-tat.

Thus, starting with N gives M an expected payoff $12p$. Starting with G gives to M an expected payoff $19p - 5(1-p)$. Hence, M starts with G if $24p - 5 > 12p$, i.e., $p > 5/12$.

As a side note, this is also an SPE since both types of J clearly play their best responses. M plays an NE in the whole game and also the last stage is played as a best response.

A more methodological remark is that in this specific game, the presence of a Tit-for-Tat type for J that initially seeks for collaboration, obtains that M reciprocates. Actually the one from M's side is not a heartfelt collaboration, rather just an attempt to make the most out of the game, but this characteristic is shared by the standard case of the repeated prisoner's dilemma. Also, it is required that p is sufficiently high for this to happen. The idea is that selfish M wants to take advantage of the good behavior of J's Tit-for-Tat type. M is smart and realizes that, if defecting immediately, both types of J will be selfish in round 2. But she can keep the Tit-for-Tat type on a leash if she collaborates at round 1, and then J will bring gifts on both occasions (Christmas and Valentine).

On the other hand, J does not intend to collaborate if his type is strategic. Yet, if the horizon of the game is further prolonged, we observe that J also switches to collaboration in the first round(s) when his type is strategic. In other words, the presence of a Tit-for-Tat type (acting as a rewarding/punishing entity) plays the same role of the carrot-stick options in the analysis of multi-stage games. Of course the likelihood of this type must be sufficiently high (so we require that p is above a threshold) in the same manner as the discount factor must be sufficiently close to 1 in Friedman's theorem, for this punishment to be effective/credible.

Exercise 6.10. A bank is being robbed by a lone bandit (B) showing the clerk (C) what seems to be a fake gun. The clerk is considering to either punch the bandit (P) or stay put (S). In the latter case, the game ends with the bandit getting away with the money, but nothing happens to the clerk besides the nasty experience. Thus, the payoffs are $+100$ for the bandit and 0 for the clerk. If C decides to intervene and punches the bandit, B can either react (R) or flee (F). The outcome actually depends on whether the gun is fake or not. If the gun is fake, playing R by the bandit just causes a fistfight, after which the bandit flees anyways; both players end up with some bruises and a payoff of -10. If the gun is real, and the bandit reacts, C gets shot, which is a very negative outcome implying payoff of -100 for him; also, the bandit robs the bank anyways, but the payoff is just $+80$ because the shooting is not pleasant. Fleeing instead causes a payoff of 0 for the bandit, while C gets $+10$ because of the satisfaction having averted the robbery. While the bandit obviously knows whether the gun is fake or not, the clerk has only an estimate of the probability p of the gun being fake.

1. Describe this game in extensive form.
2. Considering a type-agent representation of player B, draw the normal form of the game.
3. Find the Bayesian Nash equilibria in pure strategies as a function of p, and discuss them.

Solution

1. This is a Bayesian entry game. The extensive form of the game is the tree depicted in Figure 6.12.

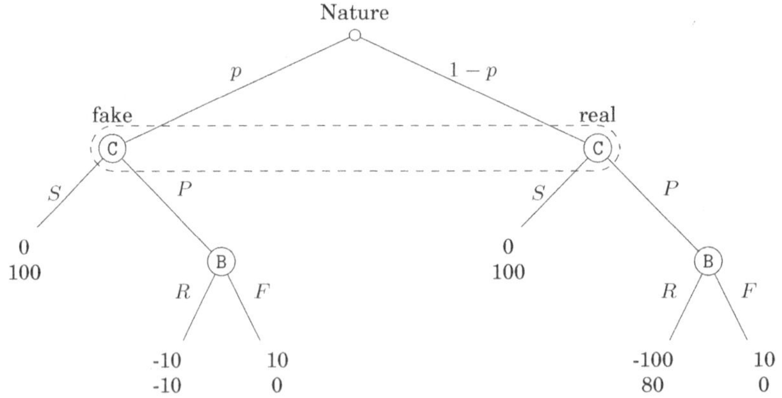

Figure 6.12: Extensive form of the Bayesian game.

2. Type-agent representation of player B decides a strategy as a pair of actions: what to do with a fake gun and what to do with a real gun. Both elements

can be either R or F. For example, RF means that B reacts only if the gun is fake, and while he flees if the gun is real.

B

C		RR	RF	FR	FF
	P	$90p$-100, 80-$90p$	10-$20p$, -$10p$	$110p$-100, 80-$80p$	10, 0
	S	0, 100	0, 100	0, 100	0, 100

3. The best response of B to P is to play FR, and all the strategies are best response to S.

   For C, the best response to RR and FF are always S and P, respectively. The best response to RF is P if $10 - 20p \geq 0$, that is for $p \leq 1/2$. In the same way, P is best response to FR if $110p - 100 \geq 0$ so $p \geq 10/11$.

   Since this is a screening game, subgame-perfect Nash equilibria are the appropriate equilibrium concept. The game can have up to three Bayesian Nash equilibria:

   - The first is (S, RR), whatever the value of p. This is not subgame-perfect, since it contains non credible behavior – i.e., B reacting when the gun is fake – that never comes into play.
   - The second is (S, RF) for $p \geq 1/2$. Also in this case B plays R with a fake gun, so the equilibrium is not an SPE.
   - The last one is either (P, FR) for $p \geq 10/11$, or (S, FR) for $p \leq 10/11$. This is also an SPE, since the behavior of the players is rational both when the gun is fake and when it is real.

Note that the two non-subgame-perfect BNEs describe a non-credible threat by B, which is to react even when the gun is fake. This is clearly an NE, since, if this was true, it would be better for C to stay put in any case. But it is not a subgame-perfect equilibrium because B does not choose an NE in the subgames in which C reacts. As for the SPE, instead, this is the logical conclusion of the game and you can understand that a really high value of p must be required for C to play the hero, otherwise C will stay put.

Exercise 6.11. Emma (E) and Frank (F) are a pair and have been invited to a wedding. They decide what to wear independently and unbeknownst to each other. Emma can choose either an aquamarine (A) or a beige (B) dress, while Frank can choose either an azure (A) or brown (B) suit. It would be best to go to the wedding with matching colors (either azure and aquamarine; or brown and beige). However, while Emma's dresses are always ready, Frank is a bit clumsy when doing laundry and may have screwed up washing and ironing his suits. Upon opening his closet, Frank may find that both his suits are not properly clean with probability p, and they are both okay with probability $1 - p$. Frank realizes this before selecting his suit; Emma does not know, she just knows the value of p (it is a common prior). If Frank's suit is properly clean, the final payoff for the pair is the same: 3 for both if they go with the A combo, 2 for both if they go with the B combo. If they go with unmatched colors (but at least Frank' suit is clean) they both get payoff 0. If F goes to the wedding with a brown suit in a bad shape, both players loses 1 from their corresponding payoffs of the "clean" case. If F wears an azure suit in a bad shape, their payoff is also decreased, but F's is decreased by 3 and E's by 5.

1. Represent this Bayesian game with E and F in extensive form.
2. Represent this Bayesian game in normal form, with a type-agent representation of player F.
3. Find the Bayesian equilibria of the game if $p = 0.8$.

Solution

1. The extensive form is the tree depicted in Figure 6.13.

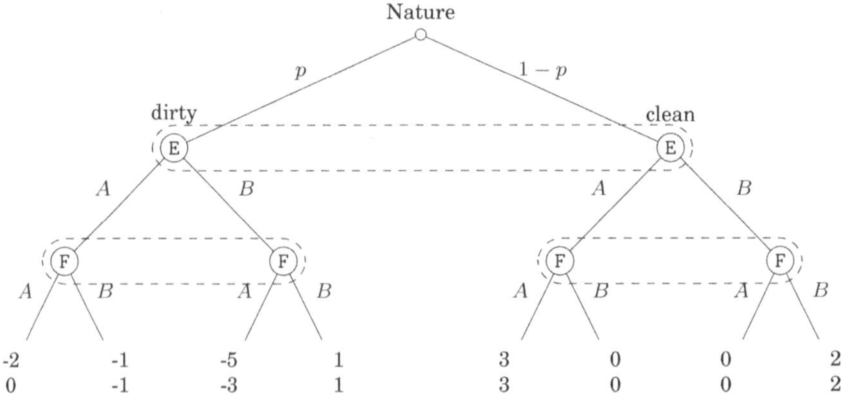

Figure 6.13: Extensive form of the Bayesian game.

The root node is Nature's move about the clothes being dirty or clean with p and $1 - p$. Below it, E moves without distinguishing between the two nodes and has two alternatives, A or B. Below that, F moves, with full knowledge about his type, choosing between A and B.

2. Type-agent representation of player F decides a strategy as a pair of actions: what to do when the clothes are dirty, and what to do when the clothes are clean. Both elements can be either A or B. For example, AB means F chooses A if dirty, B if clean. This leads to the following bi-matrix.

F

E		AA	AB	BA	BB
	A	3-5p, 3-3p	-2p, 0	3-4p, 3-4p	-p, -p
	B	-5p, -3p	2-7p, 2-5p	p, p	2-p, 2-p

3. If $p = 0.8$ the matrix above becomes

F

E		AA	AB	BA	BB
	A	-1, 0.6	-1.6, 0	-0.2, -0.2	-0.8, -0.8
	B	-4, -2.4	-3.6, -2	0.8, 0.8	1.2, 1.2

The only Bayesian Nash equilibrium is (B,BB). The intuition behind it is that it is very likely that Frank is clumsy and the clothes are dirty. Thus the azure suit is unlikely. Emma anticipates this and plays B. Thus, it is best for Frank to match this and play B too. Since this is a static game, the BNE is a sufficient equilibrium concept. However, it can also be framed as a Perfect Bayesian equilibrium by considering that the game has only one subgame and that the system of beliefs is still the prior, i.e. $\mu = 0.8$ (at the time of her play, E does not learn any further information about the state of F's clothes).

Exercise 6.12. The Alpha company (A) launches a new gadget in a market dominated by Bad Guys Inc. (B). This gadget can be *effective* or *useless*: the probability of either is 0.5 and this is common knowledge. However, company A is actually aware whether the gadget is effective or not; further, they can choose to indict a huge promotional campaign (C) or to keep a low profile (**L**). Choosing C costs 20 to A and has the effect to reveal to everyone (including the incumbent competitor B) whether the gadget is effective or not, while **L** keeps this hidden. After this decision made by A, it is now B's turn to decide whether to fight (F) or yield (Y) this new gadget. The payoff are computed as follows. If B yields, both companies earn revenue 50. If B fights, companies A and B get respective revenues of 80 and 0 if the gadget is effective; if the gadget is useless, these values are swapped. The utility of B is its revenue, while A gets the revenue minus the possible cost of the campaign, if they have indicted it.

1. Give the extensive form representation of this game.
2. Give the normal form representation of this game. Clearly state what do you consider as A's and B's strategies.
3. Find the perfect Bayesian equilibria, classify whether they are pooling, separating, or hybrid, and give their systems of beliefs.

Solution

1. This is a Bayesian game, most specifically a signaling game, where A is a Bayesian player, and the effectiveness of the gadget is A's type. The extensive form is the tree depicted in Figure 6.14.

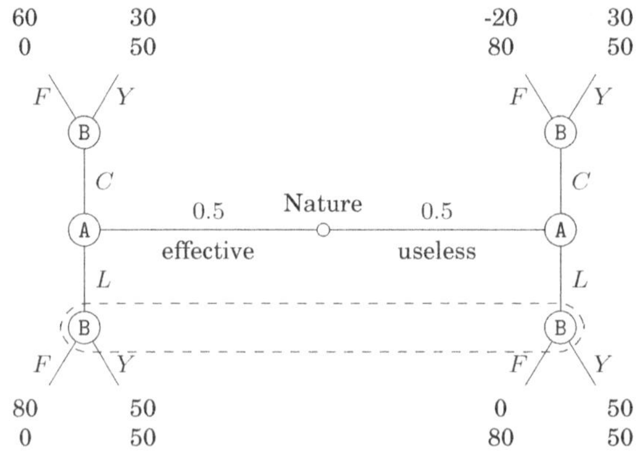

Figure 6.14: Extensive form of the Bayesian game.

Nature moves first, revealing its own type to player A. A can then move and decide whether to play C or L. The former is revealing the type to B too, while the latter is not, so the two nodes below are joined by a dashed line.

At the resulting four nodes, B can play F or Y, and payoffs are computed in the 8 leaf nodes.

Note that, as C is a revealing action, the upper part can be simplified thanks to sequential rationality, since B will always play Y and F, respectively, in his upper nodes. The upper part actually only has one rational choice by B, and the payoffs are $30, 50$ when effective (B always yields) and $-20, 80$ when useless (B always fights). The resulting tree is shown in Figure 6.15.

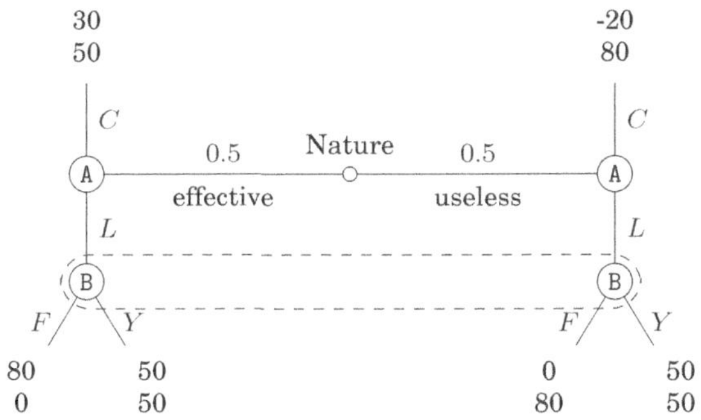

Figure 6.15: Simplified extensive form of the game after applying sequential rationality.

2. The normal form representation can then simplify B's strategy since it makes sense to consider only a binary choice of F or Y when A plays L. Every other choice by B is automatic. A's strategies are pairs of C/L values describing what to do when the gadget is effective and when it is useless. For example, CC means that A launches the revealing campaign no matter what (pooling strategy) while CL means they only do when the gadget is effective (separating). Conversely, B's strategies are what to do when A plays L, which is a single binary choice between F and Y.

   The resulting bi-matrix is then:

A \ B	F	Y
CC	5, 65	5, 65
CL	15, 65	40, 50
LC	30, 40	15, 65
LL	40, 40	50, 50

   LL is a strictly dominant strategy for A and the only NE is (LL, Y). This induces a pooling PBE if we specify the beliefs of B in the information sets that are not singletons. Since it is a pooling PBE, the beliefs of B after seeing L are just the prior. Also note that the full strategy of B also dictates that if A plays C and reveals the type, the reaction is to fight when useless, yield when effective.

Exercise 6.13. Alberto (A) and Gianluca (G) are two friends studying in the same program at the university. They have been looking for destinations to apply in the Erasmus program, and they identified two suitable ones: Coimbra (C) and Santander (S). These destinations do not receive many requests and therefore the two friends are sure that their application is surely accepted in either of them. However, they have to apply for Erasmus independently and separately, choosing one destination only (actually, they choose both, but assigning a priority, so only the first choice matters, as that application is always accepted). During their last chat, G revealed that he slightly prefers C, but his highest preference is to share the experience with his friend A. In numerical terms, this means that he assigns utility 2 and 1 to go with A to C and S, respectively; the utility in going to a different destination than A is 0 anyways. The answer of A was that he also feels the same, but G is not fully convinced: he knows that A met a female exchange student from Santander at a party and maybe he wants to go to S to meet her again. The probability that A is somehow still after the girl is p, and this value is a common prior. Of course A knows whether he is interested in the girl or not. If A wants to go after the girl, playing S is a dominant choice always giving him utility 2, while C gives 0. If A is not really interested in the girl, his utility is the same as G's.

1. Write down the extensive form representation of this game.
2. Write down the normal form of a type-agent representation of this Bayesian game (where all typed players have strategies represented as n-tuples of actions, one per type).
3. Discuss the Bayesian Nash equilibria of this game. You should find that there always is one BNE in pure strategies where A plays a pooling strategy. Find for which values of p there is another NE with a fully-separating strategy.

Solution

1. The extensive form is the tree depicted in Figure 6.16. The first node is Nature, deciding whether A is interested or not, with probability p and $1-p$ respectively. Below that, A moves and he does distinguish between the two nodes below, since he knows whether he is interested in the girl or not. His choice is either C or S, resulting in four nodes. Below that, G moves without distinguishing any of them, so he has a single information set comprising all four nodes. His choice is also C or S for any of these nodes.
2. Player A has a type, so his type-agent representation has 4 strategies (two types, two actions): CC, CS, SC, SS. The normal form representation is

		G			
		CC	CS	SC	SS
A	C	2, 2-2p	0, 2p	2, 2-2p	2p, 0
	S	0, 0	1-p, 1-p	2p, p	1+p, 1

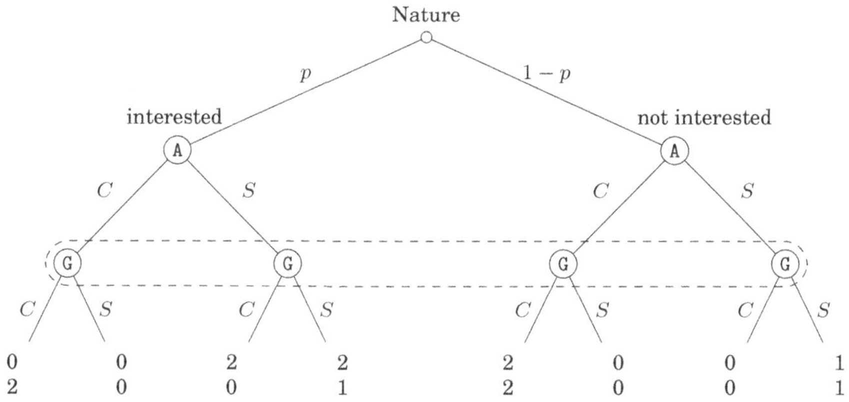

Figure 6.16: Extensive form of the Bayesian game.

3. If A plays SS, the best response for G is to play S, and vice versa (this happens since it always holds that $2p \leq 1+p$). Thus, (SS,S) is a BNE where A plays a pooling strategy, meaning that he plays the same move regardless of his type. This is the only possible BNE where G plays S, since SS is the only best response. Thus, another BNE can only be when G plays C. Indeed, if G plays C, A's best response is to play SC; however, it is not certain that C is also the best response to SC; this only happens if p is low enough (G plays C in the hope that A is not in love with the girl), whereas if p is very high, the best response for G to SC is to play S, since it is very likely that A will play S. From the matrix, we see that we need $2 - 2p \geq p$ for this condition to hold, which translates into $p \leq 2/3$, in which case (SC,C) is also a BNE. This strategy is separating, since what is played by A depends on his type.

Exercise 6.14. Jane's bank account contains €10000. Jane (J) can keep (K) the whole sum in the bank account, or buy a utility car (C) that costs €8000. The bank B is also a player, who plays simultaneously to Jane, with options being either (L) to leave Jane's money untouched in the account, or (S) to buy some subprime bonds for Jane, investing every euro that Jane has left in the account (either 2000 or 10000). After one year, Jane's account is increased with some interest $\mathcal{G}$. If the bank played L, $\mathcal{G}$ is equal to 5% of the initial amount. In case the bank played S, $\mathcal{G}$ depends on whether the bank is a "good" bank or a "bad" bank. Jane estimates the probability of "good" as p and this is a common prior. However, the bank is also fully aware of its type. A good bank playing S gives Jane her money back, plus $\mathcal{G} = 15\%$ of yearly interest rate. A bad bank steals every euro that Jane put in the bank account ($\mathcal{G} = -100\%$). The final payoff are computed after 1 year as:
— **for Jane**: the sum of the money in the account (i.e., the initial value increased by $\mathcal{G}$) plus the value of the car, if she owns one: a one-year old car is worth €5000.
— **for the bank**: if $\mathcal{G}$ is positive, then their payoff is $\mathcal{G}$; otherwise it is $-\mathcal{G}/5$.

1. Represent this Bayesian game with J and B in extensive form.
2. Represent this Bayesian game in normal form, with a type-agent representation of player B (i.e., its strategies are n-tuples of actions, one per type).
3. Find the Bayesian equilibria of the game if $p = 0.2$ and discuss them.

Solution

1. The extensive form is the tree represented in Figure 6.17.

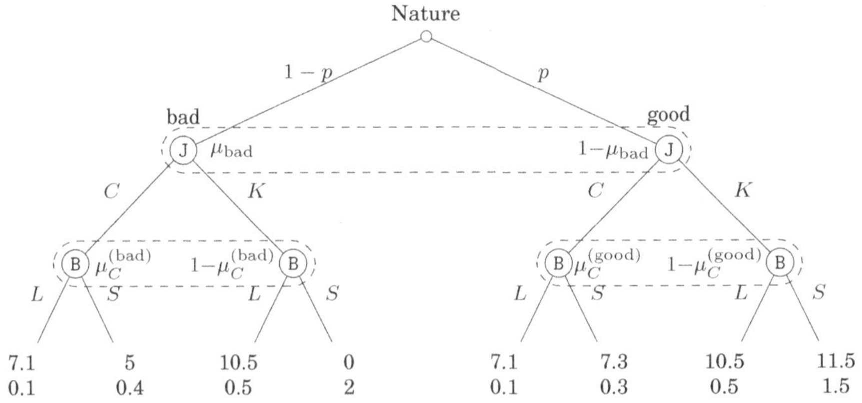

Figure 6.17: Extensive form of the Bayesian game.

All the values are reported in k€, that is, a value of 1 in the tree represents €1000. By the way, if such a rescaling by a factor 1000 is applied to all utility values, it does not affect the Nash equilibria that are unchanged.

The root node is Nature's move about the bank being good or bad with probabilities p and $1-p$. Below it, J moves without distinguishing between the two nodes and has two alternatives, K and C. This result in 4 nodes where the bank moves, always choosing between L and S, and doing so with full awareness about their type but without knowing what J chose as her move, which is why nodes are grouped like that.

2. Type-agent representation of player B decides a strategy as a pair of actions: what to do when good, and what to do when bad. Both elements can be either L or S. For example, LL means the bank always chooses L. LS means B chooses L if good, S if bad. This leads to the bi-matrix:

		B			
		LL	LS	SL	SS
J	K	10.5, 0.5	10.5p, 2-1.5p	10.5+p, 0.5+p	11.5p, 2-0.5p
	C	7.1, 0.1	5+2.1p, 0.4-0.3p	7.1+0.2p, 0.1+0.2p	5+2.3p, 0.4-0.1p

3. If $p = 0.2$ the matrix above becomes

		B			
		LL	LS	SL	SS
J	K	10.5, 0.5	2.1, 1.7	10.7, 0.7	2.3, 1.9
	C	7.1, 0.1	5.42, 0.34	7.14,0.14	5.46, 0.38

Since SS is a dominant strategy for B, the only (Bayesian) NE is (C,SS). For this simultaneous-move Bayesian game, it is enough to consider the BNE as the equilibrium concept. Yet, if one wants more advanced definitions, we see that, since the game only has one NE, it must also be subgame-perfect. And, if a proper system of beliefs is included, it is also a PBE.

Including a system of beliefs is tedious but not difficult for this case. There are 3 information sets in the tree of Figure 6.17, and they all comprise two nodes. So, as done in the figure, we introduce a belief for J that tells whether she thinks that the bank is bad (left node, belief μ_{bad}) or good (right node, belief $1-\mu_{\text{bad}}$). J cannot observe the move of B, because it is simultaneous, and she anticipates that B will play a pooling strategy, so her belief can only be the prior, i.e., $\mu_{\text{bad}} = p$. The bank instead knows its own type, but cannot observe J's move beforehand – although written above in the tree, it is actually simultaneous. So we introduce two pairs of beliefs (the second value is always 1 minus the other), namely, $\mu_C^{(\text{bad})}$ on the left side and $\mu_C^{(\text{good})}$ on the right side, that represent the respective beliefs of the bad bank and the good bank that J will play C. However, B knows that J plays C at this BNE, so we get $\mu_C^{(\text{bad})} = \mu_C^{(\text{good})} = 1$.

Side note: this lengthy analysis of the PBE confirms that this equilibrium concept is redundant when the game is so simple that the standard BNE is enough. Indeed, the PBE requires extra effort to express the beliefs, but in this specific case they do not bring any additional value to the description.

Exercise 6.15. Lorenzo (L) is a student of a Game Theory course, for which a project must be submitted. The project will be graded from 0 (worst) to 8 (best) and can be done by a student alone or a group of two. Lorenzo plans to invest effort $e \in \{0, 1, 2, 3, 4\}$ on the project and wants to ask his classmate Maria (M) to join him. Lorenzo knows that if he manages to work with Maria, their final grade for the project will be the sum of e plus Maria's contribution m that depends on how good is Maria at this subject. To convince Maria to join him, Lorenzo makes a first move revealing to her the value of his intended effort e. Then, Maria makes her move, i.e., she says whether she wants to do the project with Lorenzo or not. If she says yes, then her proficiency in Game Theory is revealed as being $m = 1$ if Maria is not very skilled in Game Theory and $m = 3$ if she is good. If she says no, they work alone on different projects, which gives grade $2e$ to Lorenzo's project. Maria's individual project gets grade 3 if she is not good at Game Theory, 6 if she is good. Maria is risk-neutral and her payoff is the grade of her project (regardless of whether it is done with Lorenzo or alone): in case two alternatives give the same payoff, she prefers working with Lorenzo. Lorenzo's payoff is his grade minus e; additionally, if he manages to do the project with Maria, his payoff will be increased by 2. All these information are common knowledge, the only thing that Lorenzo ignores is whether Maria knows Game Theory well: Lorenzo estimates this is the case with probability p, while she does not master it properly with probability $1 - p$.

1. Determine what players of this game have a type and write down the extensive form representation of this game.

2. If a type-agent representation (the strategy is an n-tuple of alternatives, one per type) of player Maria is given, how many strategies does she have? How can their number be reduced by considering only Maria's strategies with an acceptance threshold for Lorenzo's offer?

3. Assume the prior p of Lorenzo's estimate about whether Maria has good knowledge of Game Theory is strictly between 0 and 1. What is the Bayesian Nash equilibrium of this game?

Solution

1. Players are L and M and only M has a type, which can be either "good" or "bad". The extensive form is a tree, shown in Figure 6.18, where first move is Nature's determining M's type (good with probability p, bad with probability $1 - p$). Then, first mover is L with 5 alternatives (the possible values of e). Finally, M says yes (Y) or no (N).

2. In principle, M has two types and she can say Y or N to any of the 5 offers made by L. So, there are 32 possible strategies that a type can play, and overall there are $32^2 = 1024$ possible strategies of the type-agent M. However, it is sensible for M to play strategies that are of the form: "Play Y if $e \geq \theta$, N otherwise," where θ is a threshold value in L's offer. There are

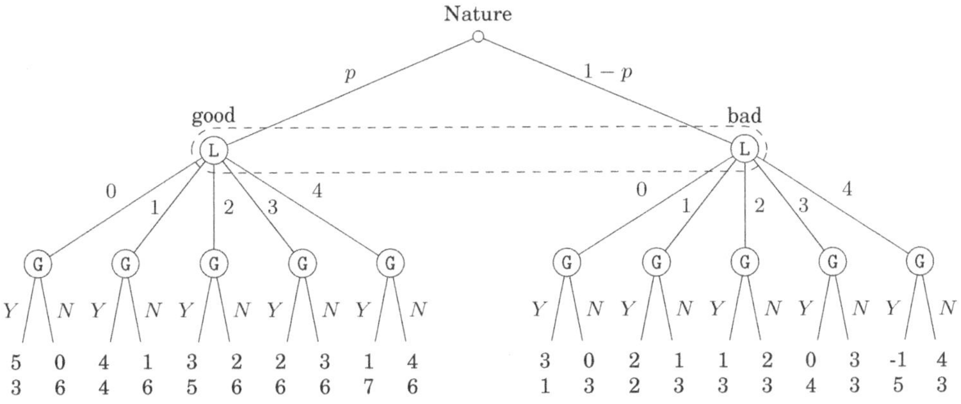

Figure 6.18: Extensive form of the Bayesian game.

6 possible such strategies (from always say N if $\theta = \infty$ to always say Y if $\theta = 0$. In this way, there are 36 possible strategies of the type-agent M.

3. M will accept an offer of e if her return is as high as her other option, doing the project alone. This means that a "good" M will accept $e \geq 3$, since she will get 6 (same as the other option), while a "bad" M will be satisfied even by $e = 2$. Thus, $e \leq 1$ will be always rejected by M. L does the project alone getting payoff equal to e. If $e \geq 3$, the offer will be accepted by both types of M, yielding a utility for L of

$$p(5 - e) + (1 - p)(3 - e) = 2p + 3 - e \leq 2p,$$

while $e = 2$ yields

$$p \cdot 2 + (1 - p) \cdot 1 = p + 1.$$

Since $0 < p < 1$ the last choice is the best one for L. Thus, L will offer $e = 2$ that is accepted by bad M and rejected by good M.

Exercise 6.16. A student (S) wants to attend a seminar by a professor (P). The seminar is held in the Aula Magna, but can also be attended online. Because of an ongoing pandemic, P is supposed to check a green pass for whomever enters the Aula Magna, and students are supposed to have one. So the first rational choice is made by S who can display (D) their green pass, or keep it hidden (H). Whatever this choice, P can decide to admit (A) the student to the Aula Magna, or keep S out and tell them to go attend online (O). To complicate things, P knows that some students have a fake green pass, and the probability of S's green pass being fake is $p = 0.1$. S clearly knows if their green pass is authentic or not. The fake green passes look real, so P has no way to tell a fake green pass apart from a valid one. The payoffs are eventually computed as follows. Case 1: S holds a valid green pass. If the student is admitted, both S and P get $+10$. If S shows their pass and P does not admit them, P gets -50. Every other payoff of either S or P is 0. Case 2: S holds a fake green pass. Then both players get a negative payoff if the student is admitted. This is -100 for both if S showed the green pass, while if the student is admitted without showing the (fake) green pass, their payoffs are -50 for S and -200 for P, respectively. If S shows a fake pass but is eventually not admitted, S's payoff is -200. Every other payoff of either S or P is 0. All of these values, including the prior, are common knowledge among the players.

1. Represent the game in extensive form.
2. Represent the game in normal form with a type-agent representation of S.
3. Find the PBEs in pure strategies and discuss whether they are separating or pooling.

Solution

1. The extensive form is a tree, shown in Figure 6.19.

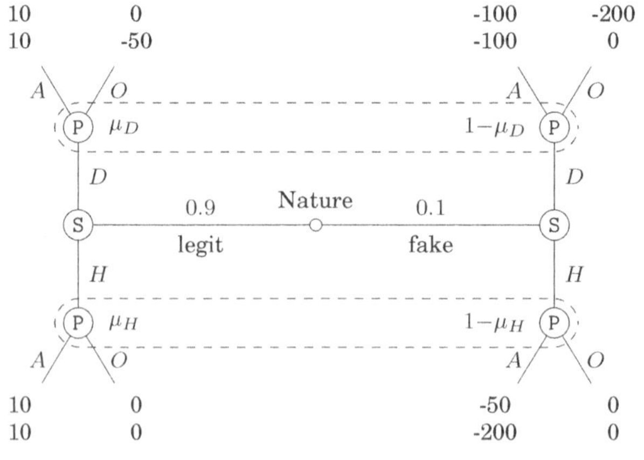

Figure 6.19: Extensive form of the Bayesian game.

P

		AA	AO	OA	OO
S	DD	-1, -1	-1, -1	-10, -45	-20, -45
	DH	4, -11	9, 9	-5, -65	0, -45
	HD	-1, -1	-11, 9	-10, -10	-20, 0
	HH	4, -11	0, 0	-5, -20	0, 0

At the root, Nature decides whether S's green pass is legit (probability $1-p = 0.9$) or fake (probability $p = 0.1$). Afterwards, S chooses D or H. Then P moves, without being able to distinguish S' types. Note that D is *not* a revealing action, because the professor is said to be unable to tell a fake green pass apart from a legit one.

2. The normal form must consider 4 strategies of S, being pairs of binary D/H values, representing "what to do if green pass is legit/what to do if fake." For example, DH means that S displays it when legit, and hides when it is fake. For player P, there are also four strategies, but now it is pair of A/O values that reflects the reaction to D and the reaction to H. For example OA means that the student is kept out when displaying the green pass, and admitted if it keeps it hidden. The result is the following bi-matrix.

3. There are two Nash equilibria: (DH, AO) and (HH, OO). The former is separating and the latter is pooling. Note that the request about the separating equilibrium is justified in that there are no revealing actions, since as discussed above, even if the student shows the green pass, it can be fake, and the professor would not know; and of course, if the student keeps it hidden, the professor would be even more in the dark! Yet, if equilibrium (DH, AO) is played, the type is revealed, since the student's strategy corresponds to displaying the green pass only if it is legitimate, and keeping it hidden tacitly acknowledge it is fake. To this, the professor replies by "following the signal," i.e., admitting in the Aula Magna only the students who display the green pass (which means it is legitimate) and sending online those who do not. However, there is also another BNE (HH, OO) that is pooling, in which nobody shows the green pass, and the professor sends every student online.

   To frame a BNE as a PBE we must also set the beliefs of P, which is done in Figure 6.19 by setting probability values μ_D and μ_H for the authenticity of the green pass (and corresponding complementary values μ_D and μ_H for that it is fake) in the respective information sets following a choice of D or H, respectively, by S.

   For the separating BNE (DH, AO) this is very easy. Naturally, $\mu_D = 1$ and $\mu_H = 0$, which directly extends the BNE to a PBE.

   For the pooling PBE (HH, OO), things are more complicated. First of all, the belief μ_H of P after seeing H follows the prior, since no extra information is gained. Thus, $\mu_H = p$. For what concerns μ_D, since D is never played, that belief can be arbitrary.

However, what is the rational best response played by P in response to D? This question is very hypothetical since D is never played, still we want to see whether the result is consistent with the rationality of the players.

In response to D, P can get an expected payoff of $10\mu_D - 100(1 - \mu_D)$ when admitting and $-50\mu_D$ when keeping out, so it turns out that depending on the value of μ_D there are two different set of PBEs: (1) the choice of P is A if $\mu \geq 5/8$ and (2) the choice of P is O if $\mu \leq 5/8$.

Exercise 6.17. The sheriff (S) is watching over the ranch, because he has been informed that a cattle thief (T) is planning to break and enter. The night is very cold though, so S is considering whether to patrol (P) the area or quit (Q) and go home; after all, T may have already fled, scared by the possibility of being caught. Indeed, T has two available options: to enter (E) the ranch to rob it or to flee (F). Clearly, both players make their decision without consulting each other first, so they can be considered to play simultaneously. One relevant aspect for T is whether S is armed or not: it is known by everybody that S carries his gun only 30% of the times. But S knows for sure whether he has brought his gun with him or not. Payoffs are computed as follows. For T, fleeing the scene gives payoff 0. Also for S, playing Q always gives payoff 0, while patrolling the area over the cold night when T has fled the scene is worth -10. If T decides to enter when S has gone home, he will commit a robbery without being disturbed, and will get payoff $+10$. All of these payoff are unaffected by S being armed or disarmed, since the two never meet at the ranch. The only difference is when T enters the ranch and S decided to keep patrolling the area. In this case, if S is unarmed, there will be a fist-fight, with both players being hurt, but T eventually will manage to escape; so, in the end they both get payoffs -20. However, if S is armed, T will end up being shot and caught. So, T gets payoff -100, while S gets $+20$ for having caught the cattle thief.

1. Represent this Bayesian game with S and T in extensive form.
2. Represent this Bayesian game in normal form, with a type-agent representation of player S (i.e., its strategies are n-tuples of actions, one per type).
3. Find the Bayesian equilibria of the game.

Solution

1. The extensive form is a tree. The root node is Nature's move about S having a gun or not with respective probabilities 0.3 and 0.7. After that, players move simultaneously, so one possibility is to have T moving first without distinguishing between the two nodes, as shown in Figure 6.20. Another possibility is to have instead S moving first and knowing whether it is the left-hand or right-hand subtree, and afterwards T moves but without knowing what node he is at.
2. Type-agent representation of player S decides a strategy as a pair of actions: what to do for each of its types, sorted as armed–disarmed. Both elements can be either P or Q. This leads to the following bi-matrix.

		S			
		PP	PQ	QP	QQ
T	E	-44, -8	-23, 6	-11, -14	10, 0
	F	0, -10	0, -3	0, -7	0, 0

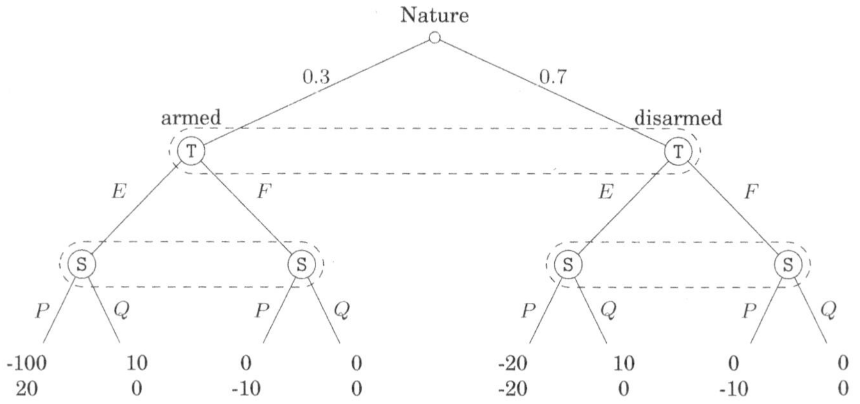

Figure 6.20: Extensive form of the Bayesian game.

3. There is no equilibrium in pure strategies. It is also clear that PP and QP are strictly dominated for S, since they always get a negative payoff, while QQ always get 0. Thus, T must mix between his only two options, and S must mix between PQ and QQ. Call ϵ the probability that T plays E, and ρ the probability that S plays PQ. The values of ϵ and ρ at the BNE can be derived as:

$$
\begin{aligned}
&9\epsilon - 3 = 0 && \Rightarrow \epsilon = 1/3 \\
&-23\rho + 10(1-\rho) = 0 && \Rightarrow \rho = 10/33
\end{aligned}
$$

Thus, the sheriff always goes home if unarmed. If armed, he patrols the area with probability $10/33$. The thief enters the ranch with $1/3$ probability.

Note: since the game is played simultaneously, this description is enough and more advanced equilibrium concepts do not add much. For example, if we want to add a system of beliefs for T, this is just the prior.

The photograph above shows a statue of Saint Cajetan Thiene in Naples, in front of the church where he is buried. Saint Cajetan (Italian: Gaetano dei Conti di Thiene, 1480-1547), was a Catholic reformer, co-founder of the Theatines. He was born in Veneto region from a noble family and graduated from the University of Padova. He was known for many acts of piety for which he is recognised as a saint in the Catholic Church. Since he founded institutions to help the poor, especially assisting those who went bankrupt because of gambling, he is considered to be the patron saint of gamblers, and therefore... of Game Theory too!

Made in the USA
Middletown, DE
05 November 2026

36607866R00093